Regulated Open Multi-Agent Systems (ROMAS)

Emilia Garcia · Adriana Giret
Vicente Botti

Regulated Open Multi-Agent Systems (ROMAS)

A Multi-Agent Approach for Designing Normative Open Systems

Springer

Emilia Garcia
Adriana Giret
Vicente Botti
Department of Computer Systems
  and Computation
Universitat Politecnica de Valencia
Valencia
Spain

ISBN 978-3-319-35621-1   ISBN 978-3-319-11572-6   (eBook)
DOI 10.1007/978-3-319-11572-6

Springer Cham Heidelberg New York Dordrecht London

Printed on acid-free paper

Springer is part of Springer Science+Business Media (www.springer.com)

Contents

Part II   ROMAS Approach

Part III   Evaluation and Case Studies

Chapter 1
Introduction

The work presented in this book deals with the problem of engineering *normative open systems* using the multi-agent paradigm. Normative open systems are understood in this work as systems in which heterogeneous and autonomous entities and institutions coexist in a complex social and legal framework that can evolve to address the different and often conflicting objectives of the many stakeholders involved. The first section of this chapter gives details about the kind of systems we deal with.

This book is focused on the analysis and design stages of the development process based on the multi-agent paradigm. Therefore, Sect. 1.2 gives a brief overview of this paradigm and shows its suitability for developing normative open systems.

Finally, the last Sect. 1.3 gives an overview of the objectives and structure of the book.

1.1 Normative Open Systems

As collaborative working and decentralization of processes increase in many domains, there is more and more demand for large-scale, flexible, and adaptive software systems to support the interactions of people and institutions distributed in heterogeneous environments. In many cases, the interacting entities are bound by rights, duties, and restrictions that influence their behavior.

Domains such as health care and electronic commerce involve autonomous institutions with their own specific social and legal contexts, where information and services are exchanged under agreed terms, with new entities (providers, patients, etc.) often joining the interactions. The defining characteristics of these systems are, therefore, that they are *open* and *regulated*. First, they are open in the sense that, dynamically at runtime, external parties can interact and become part of the system. For example, a system designed to share information between healthcare clinics should allow the participation of new clinics at runtime [61]. Second, they are *regulated* in the sense that software developed to support activities in these domains must be designed to ensure that automation does not violate any internal regulation of any party or insti-

© Springer International Publishing Switzerland 2015   1
E. Garcia et al., *Regulated Open Multi-Agent Systems (ROMAS)*,
DOI 10.1007/978-3-319-11572-6_1

tution involved, and that the rights and duties of different parties are clearly specified. Among the different ways of regulating the behavior of a software system, in this book we deal with the regulation of systems by means of norms. Since most regulated systems can be specified with norms, this assumption does not limit the applicability domain. Besides, usually in real-life systems, the restrictions on the behavior are specified in legal documents by means of norms. Consequently and common to other works [71], in this book we use the term *normative open systems* to refer to systems where heterogeneous and autonomous entities and institutions interact between them in a regulated context in order to achieve their individual and global objectives.

After reviewing applications and case studies from different domains [15, 24, 79, 87] and the literature related to the theoretical analysis of these systems [71, 86], we conclude that there are some features and challenges that are inherent to these systems independently of the domain of application. The common requirements and challenges that need to be dealt with during the analysis and design of normative open systems are presented below:

Assumption of autonomous and social behavior. A particular challenge is that these systems are composed of disparate entities and organizations that often fall under different spheres of control [61]. As a result, it is common for systems to be constructed out of many divergent subsystems. In this context, interactions can often take place between components that are managed by parties with conflicting goals, different policies, incompatible data representations, and so on.

Software organizations can represent real-world institutions in the software systems [61]. The software developed can allow new interactions between real-world entities, support existing interactions, or add new functionalities to existing systems. Moreover, when we deal with large systems with many interacting components, software organizations can be used to structure the system and create virtual subsystems that simplify the design, development, and implementation of the system. Therefore, organizations can represent a set of individual software entities that cooperate as a group to offer or demand services and resources [75]. Therefore, in order to design such systems, it is necessary to explicitly specify individual entities and institutions. The design of these entities and institutions should show the individual goals, abilities, and features of each entity, as well as the global goals, abilities, and features of each institution. Besides, the design should specify the interactions and exchanges of services and products.

Assumption of heterogeneity. Since we deal with heterogeneous and autonomous entities that can have different spheres of control and have been developed by different providers, interoperability problems may emerge. Interoperability is an issue that should be solved at implementation time. However, a design that considers this potential issue can facilitate the posterior implementation task and reduce the gap between design and implementation [61].

Some of the issues that must be solved are: (i) *Distributed Data*—the required data is spread widely across all organizations, frequently using different schemas; (ii) *Technical Interoperability*—different organizations often use different (potentially incompatible) technologies; (iii) *Process Interoperability*—different organizations often employ divergent (potentially incompatible) processes to achieve their goals; (iv) *Semantic Interoperability*—different organizations utilize different vocabularies and coding schemes, making it difficult to understand the data of others.

Moreover, as these autonomous entities and institutions often operate with a range of aims and priorities in a dynamic and changing environment, they may have to regularly update their internal processes and technology. However, it is possible that changes may take place without necessarily propagating to all other parts of the system. Interactions and interchanges of services and products should be standardized and formally described to isolate the internal characteristics of the actors of the system from their interactions with the rest of the system [15].

Regulated environment. In the real world, our behavior is restricted by a set of norms derived from the law legislation and from the regulations of the institutions that we belong to or the environments where we interact. Many software systems are also normative, which means that the behavior of their entities and institutions is bounded by rights and duties [71]. Norms provide users and members of a system with expectations about what other entities of the system will or will not do. This ensures they have confidence in the quality and correctness of what occurs in the system. Norms also avoid critical status of systems from occurring and ensure that the system follows the regulations established in a specific domain or institution.

Allowance for openness. We attempt to deal with large-scale flexible systems that can have many independent sites involved in various capacities worldwide. A common feature of large-scale systems is the expectation that more sites and entities will join the system [86]. Open systems are those that are able to interact with and integrate new entities and institutions in the system at runtime [39]. In order to interact with external entities the system must use standards of communication in order to avoid interoperability problems. Open systems should clearly specify how an external entity can be integrated into the system. The most common approach is to divide the functionality of the system into roles [15]. Then, if any external entity wants to enter in the system, it has to acquire a specific role inside the system. So once a stakeholder enters a normative system its behavior is restricted by the rights and duties of the roles it is playing. The explicit specification of these rights and duties is necessary to allow entities to reason about the consequences of acquiring a specific role.

Furthermore, little trust exists between different organizations, particularly those with conflicting goals and interests. Therefore, interchanges of services and resources between internal or external entities should be formalized [87]. The details of an agreement between two entities is completely specified at runtime, however, in regulated systems it may be necessary to specify at design time which kind of relationships are allowed and under which terms [79, 115].

A normative open system is composed of a set of entities and organizations, resources, global objectives of the system, and a normative context. Since normative open system objectives and composition can change dynamically, time is another important factor. Organizations in normative open systems are composed of a set of entities that are members of this organization, a set of global objectives of the organization, a set of resources that can be used only by the members of the organization, and a normative context. In addition to the system, the composition and objectives of an organization can change dynamically. Entities can become part of an organization or leave it at any moment. Entities in a normative open system are specified by means of their capabilities, individual objectives, personal resources, and their normative context. As is shown, a normative open system may have different normative contexts: the normative context of the system defines the norms that should be fulfilled by all the entities of the system; the normative context of an organization defines the norms that should be fulfilled by all the members of this organization; finally, the normative context of an entity defines the norms that an individual entity should fulfill regarding its requirements and properties. For example, in a normative open system that is created to share resources among universities, the normative context of the system would define the restrictions on the interactions between universities; there would be an organizational normative context for each university to bound the behavior of the members of these universities; finally, some specific entities could have its individual normative context related to special restrictions regarding its role in the university (students, directors of departments, etc.).

<normative_open_system> : : = {<entity>} {<organization>} {<resource>}

$\{< objective >\} < normative\_context > < time >$

<organization> : : = {<entity>} {<resource>} {<objective>} <normative_context>

$< time >$

<entity> : : = {<capabilities>}{<objective>} {<resource>} [<normative_context>]

<normative_context> : : = {<norm>}

In sum, we can conclude that to develop normative open systems, the supporting software should reflect the social and normative contexts of the systems and at the same time maintain its flexibility and adaptability. The software should also respect the autonomy of each entity of the system and permit their interaction despite their differences in technology. Therefore, the analysis and design of such systems could be a complex task. An incorrect or incomplete definition of a normative context could raise critical issues such as the lack of robustness, security, and privacy.

1.2 Multi-Agent Systems

Multi-agent systems (MAS) technology has emerged over the last decades as a software engineering paradigm for building complex, adaptive systems in distributed, heterogeneous environments. MAS technologies are increasingly used not only in academic environments, but also in real industrial applications. In this section, we revise the general requirements presented in the previous section and match them with MAS constructions and concepts to show the suitability of MAS for developing these systems.

Assumption of autonomous and social behavior. MAS use high-level abstraction concepts that are very close to real-life concepts such as agents and roles. Agents are computer systems that are autonomous, heterogeneous, reactive, proactive, and social [119]. Moreover, nowadays the concept of organization has become a key concept in the MAS research [54]. In organizational multi-agent systems, organizations represent institutions that exist in real life or groups of agents that interact between them in a specific environment and that can be seen from outside as a whole. These high-level abstraction concepts facilitate communication with domain experts, thereby easing things such as requirements elicitation and verification [42].

Assumption of heterogeneity. Agents and organizations in MAS are assumed to be heterogeneous [119]. The interoperability problems are solved by many MAS approaches by integrating a service-oriented approach into their architecture [38, 50]. Services standardize the interactions between heterogeneous entities without restricting the technology or the process followed to offer this functionality. Integrating agents and services thus improve flexibility, interoperability, and functionality [56]. Services offer a well-defined infrastructure and high interoperability, whereas agent technology aims to provide intelligent and social capabilities (trust, reputation, engagement, etc.) for applications. Services are a powerful interaction mechanism at implementation and also at design time. The use of services during design time helps in the specification of different levels of abstraction. Services allow to specify what an entity offers or requires separately from the internal features of this entity and how it is going to offer this functionality [80].

Regulated environment. As explained in the previous section, we deal with systems that need to bound the behavior of their entities. These restrictions on the behavior are related to system specification requirements, legal documents, and internal regulations of the institutions involved. In order to adapt MAS systems to legal and restricted environments, agents' social relationships, organizational behavior, interactions, and service interchanges are regulated [12, 71]. Some MAS methodologies, architectures, and platforms have been working on explicitly integrating the high-level abstraction of *norm* [71]. The advantage of a norm-based design approach is that there is a ready way for developers to specify these regulations explicitly in the development process, such that they become part of the design. Implementing the system in a norm-aware platform can ensure their fulfillment, even if the system has been externally implemented by different providers.

Allowance for openness. In practice, openness is enabled by a design specifying exactly how a new entity must behave in order to join the system [39]. The integration of the concept of *contract* in MAS architectures facilitates the formalization of the rights, duties, and restrictions that any entity acquires when enters the system playing a specific role [109, 110]. Contracts are flexible and expressive as they allow agents to operate with expectations of the behavior of others based on high-level behavioral commitments, and provide flexibility in how the autonomous agents fulfill their own obligations [115].

Therefore, we can conclude that MAS constructions and concepts fit with the needs of normative open systems. A normative service-oriented MAS paradigm that includes the concepts of contracts would be suitable for developing these kinds of systems independently of the domain of application. However, as is shown in Chap. 3, MAS methodologies do not completely support the development of normative open systems. The significant weaknesses are related to the identification, formalization, and verification of the different normative contexts of the system during the analysis and design of these systems.

1.3 Outline

Owing to the suitability of the multi-agent approach for development of normative open systems, this book is focused on the analysis and design of normative open systems using MAS technology. Some agent-oriented software engineering methodologies deal with the development of such systems.

In the first part of this book, we try to answer the research question: *To what extent current AOSE methodologies support the development of normative open MAS?* To answer this question, we need to previously answer another research question: *What are the requirements for developing normative open MAS?* Chapter 2 analyzes the specific features for analyzing and designing systems of this kind, and Chap. 3 analyzes how current agent approaches support these requirements.

After analyzing to what extent agent methodologies support the analysis and design of these systems, we can conclude that there are some open issues in the topic. Some of these open issues are the integration of the normative context of the system during the development process; the lack of guidelines to identify and formalize this normative context; the lack of validation and verification techniques that ensure the coherence of the final design and requirements of the system; and the coherence between the individual objectives and restrictions of each entity and the global system.

The second part of this book presents ROMAS (Regulated Open Multi-Agent Systems), a new software engineering approach for analyzing and designing normative open MAS. ROMAS is composed of a metamodel, methodology, and a development framework that allows modeling normative open systems and formally verifying part of these models. Chapter 4 presents an overview of this approach.

In ROMAS, agents, roles, and organizations are defined through a formal social structure based on a service-oriented open MAS architecture. Here, organizations represent a set of individuals and institutions that need to coordinate resources and services across institutional boundaries. In this context, agents represent individual parties that take on roles in the system; within a given organization (e.g. a company), they can both offer and consume services as part of the roles they play. Beyond this, virtual organizations can also be built to coordinate resources and services across institutional boundaries. Norms defined as permissions, obligations, and prohibitions restrict the behavior of the entities of the system. Contracts are used to formalize the relationships between entities. In our approach, we differentiate between two types of contracts: contractual agreements and social contracts. This architecture and the ROMAS modeling language is described in Chap. 5.

The ROMAS methodology defines an agent-oriented development process and provides specific guidelines for identifying and formalizing: (1) the normative context of the system, (2) the entities' communications and interchanges, and (3) both the global behavior of the system and the individual features of each entity. Chapter 6 details this methodology using the FIPA standard *Design Process Documentation Template*.

ROMAS also offers a modeling tool that supports the development of normative open systems designed using ROMAS methodology. This modeling tool integrates a model checking plug-in that allows the verification of the coherence of the normative context of a system, i.e. the coherence between the restrictions and commitments of each entity and the global specification of the system. This modeling tool is detailed in Chap. 7.

The third part of this book is focused on the evaluation of the quality and usability of our proposal. Chapter 8 analyzes the ROMAS methodology regarding its support for the analysis and design of normative open systems. We have also performed empirical evaluation of the applicability of the ROMAS methodology and tools by means of analysis and design of several case studies from different domains (e-health, manufacturing, commerce, and research). The design of such different case studies has been useful to evaluate different dimensions and uses of the ROMAS methodology. The conclusions obtained through the development of these case studies are presented in Chap. 9.

Finally, Chap. 10 summarizes the main contributions of this work and presents some future work lines.

Part I
Background

Chapter 2
Requirements for Designing Normative Open Multi-Agent Systems

In this chapter, we try to answer the research question: *Which are the common requirements for developing normative open MAS?* As is presented in Sect. 1.1, normative open systems have common features, challenges, and requirements that must be considered during their development. This section analyzes the characteristics that an agent methodology for analyzing and designing systems of this kind should have. This analysis is derived from our previous studies [54, 56, 58], related literature [12, 31, 34, 40, 75, 86], and the study of case studies from different application domains [15, 24, 61, 79, 87, 115].

In the specialized literature, there is no consensus about the terminology that must be used to specify normative open MAS. Therefore, in this section, we analyze the requirements for designing normative open systems from a semantic point of view and associate these semantics to specific terms in order to reuse them in the following sections. These specific terms are highlighted in bold.

Software methodologies are composed by the specification of design constructs, a development process, and a set of guidelines that supports or automatizes some of the development decisions. In that sense, the rest of the section is organized as follows: First, Sect. 2.1 analyzes the metamodel constructions and design abstractions that are necessary to represent systems of this kind. Second, Sect. 2.2 analyzes the support during the development process that it is necessary in order to completely analyze and formalize these systems. Finally, Sect. 2.3 analyzes how the final design of the system should be validated.

2.1 Design Abstractions

This section analyzes the design abstractions that a metamodel for modeling normative open MAS should integrate. These design abstractions are related to the common properties of systems of this kind detailed in Sect. 1.1.

© Springer International Publishing Switzerland 2015
E. Garcia et al., *Regulated Open Multi-Agent Systems (ROMAS)*,
DOI 10.1007/978-3-319-11572-6_2

Regarding the assumption of autonomous and social behavior in normative open systems, we conclude that for designing systems of this kind, it is necessary to explicitly specify individual entities (called **agents** in MAS) and **organizations**. Agents represent individual entities with their personal objectives, capabilities, and resources [119]. Organizations represent a group of agents that have a common objective or real-world institutions [54]. The explicit representation of organizations at design time is beneficial in the sense that: (1) Organizations allow to divide a large subsystem in subsystems facilitating the design, the implementation, and the maintenance of the model [75]; (2) Organizations are a high-level abstraction, very close to real life that facilitates the design and the comprehension of the clients and domain experts; (3) Organizations allow creating different contexts inside the same system (each context can have its own resources, regulations, and features) [11]. The internal structure of the organizations of the system will determine how the functionality of the system is divided between its entities, the social relationships and communications among entities and how the system interacts with its environment [73].

Regarding the assumption of heterogeneity, interactions and interchanges of services and products should be standardized and formally described in order to isolate the internal characteristics of the actors of the system from their interactions with the rest of the system [37]. Common to other works [49, 97], we propose standardizing the interchanges by means of **services**. Services standardize the interactions without restricting the technology or the process followed in order to offer this functionality [56]. Services are a powerful interaction mechanism at design and also at implementation time. The use of services during the design time helps in the specification of different levels of abstraction. Services allow to specify what an entity offers or requires from the system separately from the internal features of this entity and separately from how it is going to offer or use this functionality.

Regarding the assumption that we deal with systems in regulated environments, we conclude that the behavior of the entities and institutions in the system should be bounded by rights and duties. **Norms** provide a mechanism to explicitly represent which actions are permitted, forbidden, and obliged inside the system [88]. In that sense, norms provide users and members of a system with expectations about what other agents will do or not do. Norms provide confidence in the quality and correctness of what occurs in the system. Norms avoid critical status of the systems to occur and also try to ensure that the system follow the law regulations established in a specific domain or institution. The explicit representation of norms forces developers to analyze and consider the normative environment at design time [12]. Beyond that it allows external developers to know which behavior is expected from the software that he/she is going to implement.

The interaction of several institutions and entities from different spheres of control arise the need of specifying **different normative contexts** inside the same system [61]. First, we will need to specify the normative context of a system. It is considered to be the set of norms that regulates the behavior of each entity and the set of contracts that formalizes the relationships between entities and institutions. Second, for each institution involved in the system, we will need to specify the normative context of this institution. They are specified by the set of norms that affects the entities that are members of this institution. Third, we will need to specify the normative context of each entity. It is specified by the set of norms that directly affects the behavior of this entity in a specific moment.

Regarding the scope of a norm, tree types of norms can be considered [12]. First, the **institutional norms** are the norms that regulate the behavior inside a specific institution or group of entities. These norms are related to internal regulations of this institution in the real-world or law restrictions associated to this kind of institutions in this domain. Second, the **role norms** are the norms that any entity playing a specific role inside the system must follow. The term role is used to specify a set of functionalities inside a system. The role abstraction is very close to real-life systems. For example, every person that interacts inside a university has a role (students, teachers, directors of departments, etc.). Third, the **agent's norms** are the norms that affect only to a specific entity of the system. Every individual entity may have special rights or restrictions associated to its own design and implementation. These norms are not related to the general structure of the system or the roles that this entity plays, but with the specific features of each individual entity. For example, in a virtual market where all the clients are obliged to pay in advance, one specific client may have arrived to an agreement with the company that allow this client to pay after receiving the goods.

Norms can also specify the internal structure of the system and the social relationships between their components. This means that the social structure emerges from the norms and social relationships between entities. Moreover, the use of norms to specify the structure allows the entities to reason about the structure of their system, and allows the structure to be updated dynamically at runtime. In the literature, norms of this kind are called **structural norms** [43].

Social structure architectures imply the specification of the relationship between several entities of the system. In dynamic and flexible systems, as well as in real human societies, the specific terms of the social relationship between entities can be negotiated between the entities involved. Common to other works [40, 110], we use the abstraction of **contract templates** to specify at design time the features that any contract of a specific type should have. The use of contract templates to specify these relationships provides flexible architectures and maintains the autonomy of the system about how to implement their commitments. In this work, these kinds of contracts are called **social relationship contracts** [40]. Contracts have been used in many domains in order to formalize restrictions without compromising the autonomy of the entities. This is because contracts are expressive and flexible. They allow agents to operate with expectations of the behavior of other agents based on high-level behavioral commitments, and they provide flexibility in how the autonomous agents fulfill their own obligations [115]. Contracts also allow the negotiation of the specific terms of the engagement between a stakeholder and a role. Although contracts should be specified and negotiated at runtime, at design time contract templates should be defined in order to specify contract patterns that any contract of this type should fulfill.

Regarding the assumption of openness, the design abstractions used should be able to specify how an external entity interacts with the system and how it becomes part of the system. Using a service-oriented architecture when an external entity needs to interact with the system, it only have to follow the standard specified by the service. In the case that an external entity wants to be integrated in the system, i.e. to become part of the system by offering part of the internal functionality of the system, it has to acquire a specific role of the system. Commonly in agent literature, the internal functionality of complex systems is divided using roles [119]. A role is high-level abstraction that allows specifying system using terms close to the ones used in real-life systems. For example, in a commercial interchange, we will use the roles *client* and *provider*. The external entities that want to be integrated in the system can be heterogeneous and they can be developed outside the scope of the system. However, once a stakeholder enters in a normative system its behavior should fulfill the rights and duties of the roles that it is playing. Therefore, when an entity wants to play a specific role, it has to be informed about the rights and duties associated to this role. Moreover, flexible and dynamic systems may allow entities to negotiate at runtime how each entity is going to play each role. Therefore, the rights and duties associated to each role should be described by means of contracts. Similarly to other works [43], the contract templates that specify at design time the general terms that any entity should fulfill in order to play a specific role are called **play role contract**.

Interchanges of services and resources are formalized at runtime; however, in regulated systems it may be necessary to specify at design time which kind of relationships are allowed and under which terms. Therefore, it is necessary to specify contract templates that formalize these restrictions and may establish the interaction protocols that should be executed in order to negotiate, execute, and resolve conflicts related to these contracts. In this work, these types of contract templates are called **contractual agreements**.

2.2 Support During the Development Process

The development of normative open MAS requires complex tasks such as the integration of the individual and social perspective of the system or the integration of the system restrictions in the design of the individual entities. Therefore, software methodologies should provide a set of guidelines that simplifies or automatizes these tasks. Following we present a summary of the most important guidelines that a complete methodology for normative open MAS should provide.

One of the challenges in the design of normative open systems is determining the most suitable social structure and when the system should be structured into suborganizations [75]. Although the process of analyzing which is the most suitable organizational topology could seem to be as simple as mirroring the real-world structure, it is in fact rather complex. On the one hand, if the system supports or automates existing relationships between institutions, developers should identify and analyze these relationships in order to extract the specific requirements.

On the other hand, if the system allows new interactions, they could change the existing social structure. Since the structure of the system determines the relationships and interaction among the entities and the division of tasks between them, a bad choice in the selection of the social structure could derive problems such bottlenecks, a reduction of the productivity of the system, and an increase of the reaction time of the system [116]. Therefore, methodological guidelines that support the decision of which is the most suitable structure are necessary [41, 74].

Another challenge is the identification and formalization of the normative context of a system. In the previous section, a set of different types of norms that should be formalized at design time are introduced. These norms can be derived from: (1) the specific requirements of the system (e.g., a system in which the main goal is to increase productivity during a specific period would forbid any entity from taking a vacation during this period) [104]; (2) legal documents that formalize governmental law or internal regulations of each institution (e.g., the National Hydrological Plan, the governmental law about water right interchanges) [17]; and (3) design decisions [116]. The identification of the normative context of a system is not trivial because: (1) the description of the requirements of the system provided by domain experts might be incomplete; (2) individual entities might have their own goals that conflict with the goals of the system; (3) in systems composed of different institutions, each could have its own normative context that needs to be integrated into an overall system; and (4) legal documents are written in plain text, which means that the terminology of the domain expert and these legal documents could be different.

A poor or incomplete specification of the normative context can produce a lack of trustworthiness and robustness in the system. In open systems in which every entity could be developed by a different institution, if the rights and duties are not formally specified, an entity that tries to join a system would not know how to behave. Entities could perform actions that harm the stability of the system (e.g., in a non-monopoly system, a client could buy all the resources of one type). Therefore, specific guidelines should be added to the **requirements analysis** stage in order to identify and formalize the norms that are directly related to the requirements of the system. Also, specific guidelines for identifying the norms that should be implemented in a system derived from the **legal documents** associated to the system should be provided. This identification is a complex process because such documents are usually written in plain text and the semantic meaning of the concepts described in the legal documents and in the system design can be inconsistent.

The **structure of the system** and the **relationship between the roles and entities of the system** can be explicitly specified by means of norms and contracts. As is presented in previous sections, this explicit representation provide benefits, however, it can be a complex task in complex systems. Therefore, the methodology should provide specific guidelines that simplify and automatize the task. In that sense, the methodology should provide specific guidelines for identifying institutional, role and agent norms, as well as guidelines to formalize play role and social relationship contracts.

Another challenge is the identification of when is beneficial for the system that two entities collaborate [102]. A complete methodology should help developers in this process. Beyond that, the formalization at design time of these interchanges can be a complex task. The formalization should specify which terms of the contract are mandatory and which are forbidden. So, a complete methodology should offer specific guidelines to the **identification and formalization of contractual agreements**.

Contracts are more than a set of norms [22, 89]. The specification of **negotiation, execution and conflict resolution protocols** is also an important issue in contract-based systems [114]. These protocols should fulfill the normative context of the contract and ensure that all the terms of the contract are agreed and executed. Therefore, methodological guidelines could be very beneficial in order to avoid incoherence between the contracts' clauses and their protocols and to ensure the correctness and completeness of these protocols.

2.3 Evaluation of the Final Design

The validation of the fact that the designed system fulfills all the requirements identified in the analysis stage and the verification of the coherence of the system are common open issues in any development approach. For normative open systems, these validations and verification have even greater importance due to two specific features. First, systems of this kind integrate the global goals of the system with the individual goals of each party, where these parties are completely autonomous and their interests may conflict. It is thus crucial to help developers to verify that the combined goals of the parties are coherent and do not conflict with the global goals of the system. If any incoherence is detected, the developer should be able to determine when this issue will affect the global goals and whether it is necessary to introduce norms to avoid related problems. Second, such systems usually integrate different normative contexts from the different organizations involved, which must be coherent with the contracts defined in the system. It is necessary to ensure that each single normative context has no conflicts, and also that the composition of all the normative contexts is itself conflict-free. In this respect, an open question is how consistency and coherence of norms and contracts can be automatically checked inside an organization. Therefore, guidelines for validating that the design **fulfills the normative requirements** and for verifying the **coherence of the goals of the different parties in the system** and the **coherence of the normative context** should be offered by the methodology and integrated in the development process.

As well as verification and validation, traceability is another topic that has a special importance in normative open MAS. Requirements traceability refers to the ability to describe and follow the life of a requirement, in both forward and backward direction [23]. Traceability improves the quality of software system. It facilitates the verification and validation analysis, control of changes, as well as reuse of software systems components, and so on. The ability of following the life of a requirement associated to a norm is even more important due to the dynamicity of the normative contexts of a system. For example, in the mWater case study [59], the whole system should follow the National Hydrological Plan legislation. Without traceability, any change in this law would imply the revision of the whole system. However, if it would be possible to trace each norm individually, only the norms that had changed should be revised and only the parts of the system affected by these norms should be redesigned. Therefore, **traceability of the normative context** is a desired feature in a methodology for developing normative MAS.

Chapter 3
State of the Art

In this chapter, we attempt to answer the research question: "To what extent current AOSE methodologies support the development of normative open MAS?". This chapter summarizes the state of the art of agent methodologies' support for normative open MAS regarding the requirements described in the previous chapter. Section 3.1 offers a general overview of the state of the art. Section 3.2 compares the most developed approaches for designing systems of this kind. Finally, Sect. 3.3 concludes the chapter by discussing the open issues on the topic.

3.1 General Overview

This section provides a general overview of the state of the art following the same structure used in the previous chapter to describe the requirements for analyzing and designing normative open MAS.

3.1.1 Design Abstractions

The representation of individual entities and the social structure of the system is a common topic in MAS. The concept of **organization** has become a key concept in MAS research, as its properties can provide significant advantages when developing agent-based software, allowing more complex system designs to be built with a reduced set of simple abstractions [77, 78]. Organizations comprise both the integration of organizational and individual perspectives and the dynamic adaptation of models to organizational and environmental changes. Relevant organizational methodologies are: Gaia [123], AML [113], AGR [47], AGRE [48], MOISE [66], INGENIAS [95], OperA [40], OMNI [44], OMACS [32]. A detailed survey of organizational approaches to agent systems can be found in [78].

© Springer International Publishing Switzerland 2015
E. Garcia et al., *Regulated Open Multi-Agent Systems (ROMAS)*,
DOI 10.1007/978-3-319-11572-6_3

Many AOSE approaches deal with the challenge of communicating heterogeneous entities avoiding interoperability issues by means of integrating service-oriented architectures into their architectures [56]. A service-oriented open MAS (SOMAS) is a multi-agent system in which the computing model is based on well-defined, open, loosely coupled service interfaces such as **web services**. Such services can support several applications including heterogeneous information management; scientific computing with large, dynamically reconfigurable resources; mobile computing; pervasive computing; etc. Relevant SOMAS proposals are: Tropos [26], Alive [38], GORMAS [6], INGENIAS [49].

Agents that join an organization usually have to deal with constraints, such as the need to play particular roles so as to participate in certain allowed interactions. The specification of explicit **norms** has been employed for keeping agents from unexpected or undesirable behavior [70]. Currently, the most developed agent methodologies integrate norms into their metamodels in order to formalize the restrictions on the behavior of the actors of the systems [8, 13, 33, 44]. Many of them also allow the specification of organizational systems. These agent methodologies are able to describe different normative contexts by means of specifying norms whose scope is limited to one specific organization of the system [22, 43, 109].

Another high-level abstraction construction that is becoming increasingly important for agent behavior regulation is the explicit specification of electronic **contracts** [87, 89]. Most of the approaches integrate contracts to specify the contractual agreements between parties [22, 69]. Only few approaches use contracts to specify the structure of the system and the social relationship among the system's entities [21, 43, 92].

3.1.2 Support During the Development Process

Selecting the most suitable organizational topology and distributing the functionality of the system in the most appropriate way can be a complex task in large and heterogeneous systems. Beyond the complexity of the task, bad selection of the structure of the organization can be critical to the success of the system [75]. Some MAS methodologies provide specific guidelines [9, 34]. The social structure and coordination are usually represented in agent approaches by means of roles and structured organizations. Only a small subset considers the normative context when selecting the organizational structure and only few approaches represent the social structure by means of norms in order to allow entities to dynamically reason and change this structure at runtime [40].

Few methodologies consider services as an important part of the analysis and design of the system and provide guidelines for specifying their interface as well as their internal functionality [9, 50]. Without these guidelines the designer should rely only on his/her expertise to specify the services and their attributes. This task could be very complex in dynamic, distributed, large systems.

Although some methodologies include into their metamodel and development process the description of the normative context of a system, only few works provide guidelines to actually identify the normative context of the system. Work by Boella and Rotolo et al. [12, 99] offers several guidelines that point the attention of the system designer to important issues when developing a normative system, but they cannot be used as an artifact for designers to identify the norms that regulate the system. Kollingbaum et al. [82] present a framework called Requirement-driven Contracting (RdC) for automatically deriving executable norms from requirements and associated relevant information, but this framework only derives system norms from the description of the goals of the system. A more complete guideline that includes the analysis of each entity's goals, and the resources and relationship between entities is still needed.

Breaux et al. [17, 19] present a methodology for extracting and prioritizing rights and obligations from regulations. They show how semantic models can be used to clarify ambiguities through focused elicitation, thereby balancing rights with obligations. Breaux [18] continues this work, investigating legal ambiguity and what constitutes reasonable security. This methodology identifies obligations and restrictions derived from the analysis of the complaints, agreements, and judgments of the system. It seems to address existing systems and needs runtime information to derive the norms. The methodology is not, however, focused on the analysis and design of multi-agent systems, although some of these guidelines could be combined with an agent methodology to adapt the system at runtime and increase security.

Siena et al. [104] study the problem of generating a set of requirements, which complies with a given law, for a new system. It proposes a systematic process for generating law-compliant requirements by using a taxonomy of legal concepts and a set of primitives to describe stakeholders and their strategic goals. This process must be combined with an agent methodology to completely design the system.

Saeki and Kaiya [101] propose a technique to elicit regulation-compliant requirements. In this technique, the regulations are semantically checked against requirement sentences to detect the missing obligation acts and the prohibition acts in the requirements.

3.1.3 Evaluation of the Final Design

Regarding the verification of the models and the consistency and coherence of norms and contracts inside an organization, there are some works in the literature but this is still an open problem. Most work presented in this book is focused on offline verification of norms by means of model checking [117].

The application of model-checking techniques to the verification of contract-based systems is an open research topic. Some works like [105] model contracts as a finite automata that models the behavior of the contract signatories. Other works represent contracts as Petri nets [76]. These representations are useful to verify safety and liveness properties. However, adding deontic clauses to a contract allows conditional obligations, permissions, and prohibitions to be written explicitly. Therefore, they are more suitable for complex normative systems. In [46, 94], a deontic view of contracts is specified using the *CL* language. The work in [94] uses model-checking techniques to verify the correctness of the contract and to ensure that certain properties hold. The work in [46] presents a finite trace semantics for *CL* that is augmented with deontic information as well as a process for automatic contract analysis for conflict discovery. In the context of Service-Oriented Architectures, model checkers have recently been used to verify compliance of Web service composition. In [85], a technique based on model checking is presented for verification of contract-service compositions.

In the context of verification techniques for MAS, there are some important achievements using model checking. In [118], the SPIN model checker is used to verify agent dialogs and to prove properties of specific agent protocols such as termination, liveness, and correctness. In [14], a framework for the verification of agent programs is introduced. This framework automatically translates MAS that are programmed in the logic-based agent-oriented programming language AgentSpeak into either PROMELA or Java. It then uses the SPIN and JPF model checkers to verify the resulting systems. In [120], a similar approach is presented but it is applied to an imperative programming language called MABLE. In [93], the compatibility of interaction protocols and agents' deontic constraints is verified. However, none of these approaches are suitable for many normative open systems since they do not consider organizational concepts.

There are only a few works that deal with the verification of systems that integrate organizational concepts, contracts, and normative environments. The most developed approach is presented in the context of the IST-CONTRACT project [92]. It offers contract formalization and a complete architecture. It uses the MCMAS model checker to verify contracts. However, as far as we know, it does not define the organizational normative context or verify the coherence of this context with the contracts.

Only few works ensure traceability of the requirements [23] and none of them is focused on the traceability of the normative context attributes.

As presented in this section, there are some approaches that offer partial solutions to the issues derived from the development of normative open systems. However, combination of these partial solutions to obtain a complete methodology is not an easy task. As each approach uses different terminology, semantics, and metamodel constructions, often these partial solutions are not compatible. This study also shows that as regards the requirements presented in Sect. 3.2 and among the analyzed methodologies, those that seem more suitable for normative open MAS are OperA [40], O-Mase [34], Tropos [109], and Gormas [10]. These approaches are studied in-depth in the following section.

3.2 Comparison of Methodologies

In order to analyze to what extent AOSE methodologies support the development of normative open systems, we need to analyze, evaluate, and compare the methodologies available in the literature.

Owing to the differences in terminology and semantics of each methodology, the comparison of methodologies is a complex task. The evaluation of software engineering techniques and applications is an open research topic. Some evaluation approaches are based on comparison by means of a case study [34, 35], whereas other approaches use formal techniques like model checking to assess the compliance of specific properties [14, 21]. Our approach tries to be more general. We analyze the support for the development of normative open MAS by means of a significant set of criteria related to the specific features of these kinds of systems.

Based on our previous work [54, 55, 56, 58], the study of the different approaches available in the literature [16, 25, 84, 107], and the requirements for developing these systems (Sect. 3.2), we propose a set of questionnaires that guide the analysis and comparison of methodologies for developing normative open systems. The use of questionnaires makes the answers more specific and easier to compare. It also reduces evaluation time and simplifies the evaluation process [30, 112].

The overview of the state of the art presented in the previous section shows that as regards the requirements presented in Sect. 3.2, the most developed methodologies are:

- **Organizations per Agents (OperA)** [40]: OperA is a framework for the specification of normative open MAS that includes a formal metamodel, a methodology, and a case tool. The OperA methodology is structured into three steps:

  - *Organizational model design* This phase specifies the *OperA Organizational Model* for an agent society. This model is composed of three levels: (1) Coordination Level: It specifies how the structure of the society is determined. (2) Environment Level: The society model determined in the previous step is further refined with the specification of its social structure in terms of roles, global requirements, and domain ontology. (3) Behavior Level: The organizational model of an agent society is completed with the specification of its interaction structure which results from the analysis of the interaction patterns and processes of the domain. This process is supported by a library of interaction patterns.
  - *Social model design* This phase describes the agent population in the Social Model that will enact the roles described in the structure. This phase describes the roles specified in the previous phase, role negotiation scenes, and the characteristics of the agents that apply for society roles. In other words, during this phase the social contracts that define the structure of the system are detailed.
  - *Interaction model design* This phase describes the concrete interaction scenes between agents. Interaction contracts are used to formalize these interaction scenes.

- **Organization-based Multi-agent System Engineering (O-MaSE)** [34]: provides a customizable agent-oriented methodology based on a metamodel, a set of method fragments, and a set of method construction guidelines.

  O-Mase methodology explicitly defines activities and tasks but does not define specific phases. O-Mase provides a set of guidelines to organize these activities in different ways based on project need. These activities include the analysis of the requirements; the design of the system by means of organizations and roles; the architecture design by means of defining agent classes, protocols, and policies; the low level design in which specific plans, capabilities, and actions are described; and the code generation.

- **Tropos** [20]: The initial version of the Tropos methodology was focused on supporting the agent paradigm and its associated mentalistic notions throughout the entire software development life cycle from requirements analysis to implementation [20]. Notions of agent, goal, task, and (social) dependency are used to model and analyze early and late software requirements, architectural and detailed design, and (possibly) to implement the final system. The proposed methodology spans four phases:

  - Early requirements, concerns the understanding of a problem by studying an organizational setting; the output of this phase is an organizational model which includes relevant actors, their respective goals, and interdependencies. Early requirements include two main diagrams: the actor diagram and the goal diagram. The latter is a refinement of the former with emphasis on the goals of a single actor.
  - Late requirements, where the system-to-be is described within its operational environment, along with relevant functions and qualities. The system-to-be is represented as one actor which has a number of dependencies with the other actors of the organization. These dependencies define the system's functional and nonfunctional requirements.
  - Architectural design, where the system's global architecture is defined in terms of subsystems, interconnected through data, control, and other dependencies. This phase is articulated in three steps: (1) definition of the overall architecture, (2) identification of the capabilities the actors require to fulfill their goals and plans, and (3) definition of a set of agent types and assignment to each of them one or more capabilities.
  - Detailed design, where behavior of each architectural component is defined in further detail. Each agent is specified at the micro-level. Agents' goals, beliefs, and capabilities are specified in detail, along with the interaction between them.

- **Guidelines for Organizational Multi-Agent Systems (Gormas)** [10]: GORMAS defines a set of activities for the analysis and design of organizational systems, including the design of the norms that restrict the behavior of the entities of the system.

Gormas methodology is focused on the analysis and design processes, and is composed of four phases covering the analysis and design of a MAS: The first phase is *mission analysis* that involves the analysis of the system requirements, use cases, stakeholders, and global goals of the system; the *service analysis phase* specifies the services offered by the organization to its clients, as well as its behavior, and the relationships between these services; the *organizational design phase* defines the social structure of the system, establishing the relationships and restrictions that exist in the system; finally, at the *organization dynamics design phase*, communicative processes between agents are established, as well as processes that control the acquisition of roles along with processes that enable controlling the flow of agents entering and leaving the organization. Additionally, some norms that are used to control the system are defined. Finally, the organization dynamics design phase is responsible for designing guides that establish a suitable reward system for the organization.

3.2.1 Design Abstractions

Table 3.1 shows the evaluation criteria for analyzing which design abstractions and constructions these methodologies support. These criteria are directly related to the requirements for developing normative open systems presented in Sect. 2.1.

As is shown in Table 3.2, all the studied methodologies with the exception of Tropos describe a MAS as an organizational structure. Tropos does not define explicitly

Table 3.1 Evaluation criteria: Regarding the design abstractions

Organizations: Does the methodology support the explicit representation of organizations?
Services: Does the methodology support the specification of standard services?
Normative contexts: Does the methodology support the specification of different normative contexts in the system?
Institutional norms: Does the methodology support the specification of norms that only affect the scope of a specific institution?
Role norms: Does the methodology support the specification of norms that are associated to a specific role?
Agent norms: Does the methodology support the specification of norms that are associated to a specific agent?
Structural norms: Does the methodology support the formalization of the structure of the system by means of norms?
Social relationship contract: Does the methodology support the formalization of the structure of the system by means of contracts?
Play role contract: Does the methodology support the formalization of the rights and duties that an agent acquires when it plays a specific role in the system by means of contracts?
Contractual agreements: Does the methodology support the formalization of the interchange of resources and services between different actors of the system?

Table 3.2 Design abstractions comparative

	OMASE	OPERA	TROPOS	GORMAS
Organizations	Supported	Supported	Partially supported	Supported
Services	Supported	Supported	Not supported	Supported
Normative contexts	Supported	Supported	Not supported	Supported
Institutional norms	Supported	Supported	Not supported	Supported
Role norms	Supported	Supported	Not supported	Supported
Agent norms	Supported	Not supported	Not supported	Supported
Structural norms	Not supported	Supported	Not supported	Supported
Social relationship contracts	Not supported	Supported	Partially supported	Not supported
Play role contracts	Not supported	Supported	Partially supported	Not supported
Contractual agreements	Not supported	Supported	Partially supported	Not supported

organizations, however, this methodology describes the social relationships between the entities of the system by means of dependencies. The benefits of using the abstraction of organization instead of dependencies are that organizations create different contexts, they are close to real-life institutions, and they divide the system into different subsystems facilitating the modulation of the system.

All the studied methodologies with the exception of Tropos integrate the specification of services into their metamodels.

O-Mase regulates the behavior of the entities by means of a set of norms called *policies*. These policies describe how an organization, role, or agent may or may not behave in particular situations. O-Mase does not integrate the concept of contract. In that sense, O-Mase does not support the specification of commitments between entities and does not explicitly specify if an entity can negotiate the norms or policies that assume when playing a specific role.

The OperA model regulates the behavior of the entities by means of norms and contracts. Norms specify obligations, permissions, and prohibitions of the roles of the system. OperA does not include the design of individual agents. However, it assumes that agents can understand the society ontology and communicative acts and are able to communicate with society. OperA defines two types of contracts: *social contracts* and *interaction contracts*. These abstractions, respectively, match with the *play contract* and *contractual agreement* concepts detailed in Sect. 2.1. Social contracts establish an agreement between the agent and the organization model and define the way in which the agent will fulfill its roles. In that sense, the structure of the society is defined by the social contracts specified in the system. Interaction contracts establish an agreement between agents, i.e., they define agent's partnership, and fix the way a specific interaction scene is to be played.

The initial version of the Tropos methodology [20] does not support the concepts of norms or contracts. However, Telang et al. [109] enhance Tropos with commitments. It proposes a metamodel based on commitments and a methodology for specifying a business model. The concept of commitment in Telang et al. [109] match the concept of contractual agreement used above. The specification of social contracts is not supported by this approach.

Gormas allows the specification of institutional, role, agent, and structural norms. However, Gormas does not support the concept of contract, neither to formalize social relationship, nor to specify contractual agreements.

3.2.2 Support During the Development Process

Table 3.3 shows the evaluation criteria for analyzing to what extent these methodologies offer guidelines that support the design of the specific features of the normative open systems. These criteria are directly related to the requirements for developing normative open systems presented in Sect. 2.2.

Table 3.4 shows the comparison of the selected methodologies regarding the criteria detailed in Table 3.3. The results of this analysis are described below.

Although O-Mase includes a specific task where the policies (norms) of the system are formalized, it does not provide guidelines that help the designer to identify these policies from the requirements of the system, legal documents, or systems design. This identification relies on the designer expertise.

Table 3.3 Evaluation criteria: Regarding the support during the development process

Coverage of the lifecycle: What phases of the lifecycle are covered by the methodology?
Social structure: Does the methodology provide any guideline to identify the best social structure of the system?
Requirement norms: Does the methodology provide any guideline to identify and formalize the norms of the system during the requirement analysis?
Does the methodology provide any guideline to identify which requirements should be specified as norms?
Legal documents: Does the methodology provide any guideline to identify and formalize the norms that should be implemented in the system derived from legal documents associated to the system?
System design: Does the methodology consider the normative context of the system as an important factor in the design of the system?
Structure considers norms: Is the normative context of the system analyzed before specifying its structure? Is this normative context integrated in the guideline to define the structure of the system?
Contractual agreements: Does the methodology provide any guideline to identify and formalize contractual agreements?
Contract protocols: Does the methodology provide any guideline to formalize the negotiation, execution, or conflict resolution protocol associated to each contract regarding its requirements?

Table 3.4 Support during the development process comparative

	OMASE	OPERA	TROPOS	GORMAS
Social structure	Provided	Provided	Not provided	Provided
Requirement norms	Partially provided	Partially provided	Not provided	Partially provided
Legal documents	Not provided	Not provided	Not provided	Not provided
System design	Considered	Considered	Not considered	Considered
Structure considers norms	Part of the normative system is analyzed before but is not integrated in the guideline	Part of the normative system is analyzed before but is not integrated in the guideline	Not considered	Supported
Contractual agreements	Not provided	Partially provided	Not provided	Not provided
Contract protocols	Not provided	Partially. It offers a library of patterns for interaction protocols	Not provided	Not provided

OperA offers guidelines to select the most appropriate organizational social structure and to specify interaction protocols by means of patterns. However, this methodology does not offer guidelines to capture the clauses (norms) that each contract should contain.

In the Tropos version presented by Telang et al. [109], a methodology is proposed to analyze and design the system. One of the steps of the methodology consists in identification of the contractual agreements derived from business processes. However, no other guideline related to the normative context of the system is provided.

Gormas offers a detailed guideline to select the most appropriate social structure. Norms in Gormas are presented from the early beginning of the development process, however, Gormas does not offer any specific guideline to identify the norms that restrict the system. Their identification lies on the expertise of the designer. Gormas neither offers guidelines for specifying the most appropriate interaction protocols regarding the specific requirements.

3.2.3 Evaluation of the Final Design

Table 3.5 shows the evaluation criteria for analyzing how agent methodologies support the specification of the final designs, and their validation and verification. These criteria are directly related to the requirements for developing normative open systems presented in Sect. 2.3.

Table 3.5 Evaluation criteria: Regarding the evaluation of the final design

Modeling tool: Does the methodology provide an associated modeling tool?

Code generation: Does the methodology or its associated tools provide a mechanism for automatic generation of code from the model?

Validation of the requirements: Does the methodology offer guidelines to validate that the requirements of the systems are fulfilled with the resulting design?

Verification of inconsistencies: Does the methodology offer guidelines to verify that there are no inconsistencies such as conflicts between the individual behavior of an agent and the global objectives of the system?

Tests: Does the methodology or its associated tools provide simulations or simplified system prototypes to experimentally check the behavior of the system?

Coherence of the normative context: Does the methodology offer guidelines to verify the coherence of the normative context?

Does the methodology offer guidelines to verify the coherence between the system and agent's goals and the normative context?

Traceability of the normative context: Does the methodology support traceability of the normative context?

Table 3.6 Evaluation of the final design comparative

	OMASE	OPERA	TROPOS	GORMAS
Modeling tool	Provided	Provided	Partially provided. The tool does not support norms and contracts	Provided
Code generation	Partially provided	Partially provided	Not provided	Partially provided
Validation of the requirements	Not supported	Not supported	Partially supported	Not supported
Verification of inconsistencies	Not supported	Not supported	Not supported	Not supported
Tests	Not supported	Not supported	Not supported	Not supported
Coherence of the normative context	Partial verification in the case tool	Partial verification in the case tool	Not supported	Not supported
Traceability of the normative context	Not supported	Not supported	Not supported	Not supported

Table 3.6 shows the comparison of the selected methodologies regarding the criteria detailed in Table 3.5. The results of this analysis are described below.

O-Mase methodology framework is supported by the aT^3 integrated development environment, which supports method creation and maintenance, model creation and verification, and code generation and maintenance [35]. The aT^3 verification framework allows selecting from a set of predefined rules which should be checked against the model. This fact allows verifying specific properties of the model and processes' consistency. However, to our knowledge, there is no tool for verifying the coherence of the normative context.

OperA models can be implemented using the Operetta tool [91]. Although OperA methodology does not integrate the verification of the system as a step of the methodology, the Operetta tool integrates model-checking techniques to verify the coherence of the system design. This verification includes the validation of the coherence of the normative context of the system.

The tool TAOM4E [90] supports the design of the Tropos methodology in the version presented in [20]. However, this tool is not suitable for design of normative open MAS because this version of the Tropos methodology does not support either norms, or contracts. Chopra et al. [26] deal with the verification of Tropos models. This work proposes a technique to verify that an agent can potentially achieve its objectives playing a specific role, and that an agent is potentially able to honor its commitments. However, it does not provide any guideline or technique to verify the coherence of the normative system. Tropos offers supports for requirements traceability but does not consider the normative context [23].

The EMFGormas CASE tool [53] supports the analysis and design of systems based on the Gormas methodology. Gormas does not offer tools for verification of the coherence of the system or the traceability of the normative context.

3.3 Conclusions: Open Issues in the Analysis and Design of Normative Open MAS

Considering the general study of the state of the art and the comparison of methodologies presented in the previous section, we conclude that:

- Most well-known agent methodologies integrate into their metamodels the concepts of organizations and norms. This fact allows designers to specify and formalize institutional, role, and agent norms, as well as specify different normative contexts inside the same system.
- Only few methodologies integrate the concept of contract in their metamodel. Some methodologies integrate into their metamodel the specification of contractual agreements, however, the use of structural norms and contracts to define the structure of the system is only supported by a small subset of methodologies.
- Most methodologies provide specific guidelines for selecting the most suitable organizational typology and for distributing the functionality of the system in the most appropriate way between the parties involved. However, only a small subset considers the normative context when selecting the organizational structure.
- No methodology integrates into the development process guidelines that completely support the identification of norms from the analysis of the requirements, nor from legal texts.

- Although there is some work related to validation and verification of the designed models, it is still an open problem. Verification using any development approach is important, but in normative open systems it is even more so due to the high risk of incoherence resulting from interference between different normative contexts, and between the global goals of the system and the individual goals of each party.
- Traceability of norms from requirements is not well supported by current methodologies.
- Although in the literature there are partial solutions to deal with the development of normative open MAS, there is no complete methodology that guides the development process. The combination of these partial solutions is not possible in many cases due to differences in terminology, semantics, and development processes.

Part II
ROMAS Approach

Chapter 4
ROMAS Approach Overview

As is presented in the previous chapter, current approaches for developing normative open MAS do not completely support the analysis and design of these kinds of systems. In this chapter, we present the ROMAS approach that deals with some of the open issues in this topic.

This chapter gives an overview of this approach by means of: (1) A description of the main objectives of ROMAS; (2) An introduction to the main concepts of the ROMAS architecture and metamodel; (3) An introduction to the ROMAS process lifecycle; (4) A brief description of the ROMAS background; (5) An introduction to the FIPA standard *Design Process Documentation Template* that is followed in the following chapters in order to specify the ROMAS metamodel and methodology.

4.1 ROMAS Objectives

The ROMAS approach covers the analysis and design of normative open systems by means of offering a specific multi-agent architecture and metamodel, a specific methodology, and a development framework that supports this metamodel and methodology. As shown in Fig. 4.1, this approach tries to cover all the requirements that were identified in the Chap. 2.

The main objective of the ROMAS modeling language and architecture is to offer a well-defined metamodel that allows graphically representing all the common features of the normative open systems.

The ROMAS methodology tries to contribute to the state of the art by offering a complete development process for analyzing and designing normative open MAS that includes a set of guidelines to identify, formalize, and verify the normative context of the system, as well as, that allows the traceability of the normative context

© Springer International Publishing Switzerland 2015   35
E. Garcia et al., *Regulated Open Multi-Agent Systems (ROMAS)*,
DOI 10.1007/978-3-319-11572-6_4

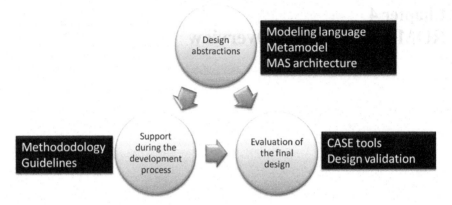

Fig. 4.1 Overview of ROMAS approach

from the requirements to the design decisions and viceversa. The general objectives of the ROMAS methodology are:

- Analyzing the system requirements from a global and individual point of view, i.e., analyzing the global requirements of the system and the individual requirements of every entity of the system.
- Analyzing and formalizing the social structure of the system and the relationships between its entities.
- Formalizing the relationships and interchanges between entities in a way that allows heterogeneous and autonomous entities to interact, even if these entities have been implemented by external providers using different technologies.
- Analyzing and formalizing the normative context of the system, i.e., the restrictions on the entities behavior derived from the system's requirements and the design decisions.
- Verifying the coherence of the designed normative context.
- Formalizing the normative context in a way that allows the traceability from the requirements to the design decisions and viceversa.

The main objective of the ROMAS development framework is to offer a graphical mechanism for modeling normative open systems following the ROMAS methodology and metamodel. It is intend to be flexible, interoperable, and modulable. Besides, another objective is to integrate the design of systems, the verification of the model, and the automatic code generation.

4.2 ROMAS Architecture and Metamodel

In ROMAS, *agents*, *roles*, and *organizations* are defined through a formal social structure based on a service-oriented open MAS architecture, whose main features are summarized in Fig. 4.2. Here, organizations represent a set of individuals and institutions that need to coordinate resources and services across institutional boundaries.

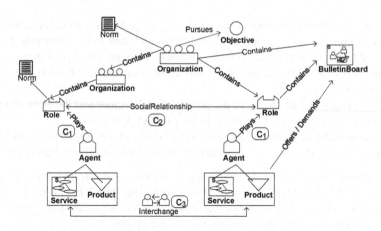

Fig. 4.2 Overview of ROMAS architecture

In this context, agents represent individual parties who take on roles in the system, within a given organization (e.g., a company), they can both offer and consume services as part of the roles they play. Beyond this, virtual organizations can also be built to coordinate resources and services across institutional boundaries. Importantly, each of these concepts must be strictly defined, alongside their interrelations. Organizations are conceived as an effective mechanism for imposing not only structural restrictions on their relationships, but also normative restrictions on their behavior. These restrictions are formalized in ROMAS by means of norms and contracts.

Norms in ROMAS are specified using the model described in [28], which defines norms that control agent behavior, the formation of groups of agents, the global goals pursued by these groups, and the relationships between entities and their environment. Specifically, it allows norms to be defined: (i) at different social levels (e.g., interaction and institutional levels); (ii) with different norm types (e.g., constitutive, regulative, and procedural); (iii) in a structured manner; and (iv) dynamically, including later derogation. Figure 4.2 shows two types of norms: (i) those that are associated with each organization; and (ii) those that are associated with each role. Clearly, the former must be complied with by any organization member, while the latter must be complied with by all agents playing that role.

Finally, ROMAS also allows interactions to be formalized by means of *contracts*. These are necessary when working in an open regulated system, to be able to specify the expected behavior of others without compromising their specific implementation. ROMAS involves two types of contracts: *social contracts* and *contractual agreements*. Social contracts can be defined as a statement of intent that regulates behavior among organizations and individuals. As shown in Fig. 4.2, social contracts are used to formalize relationships: (i) between an agent playing a role and its host organization (as indicated by the contract labeled c_1); and (ii) between two agents providing and consuming services (as indicated by c_2). Social order, thus, emerges from the negotiation of contracts about the rights and duties of participants, rather than being given in advance. In contrast, contractual agreements represent the commitments between several entities in order to formalize an interchange of services or products (c_3).

The properties of each entity of the presented architecture and the allowed relationships between them are formalized in the *ROMAS metamodel*.

In order to facilitate the modeling tasks, this unified metamodel can be instantiated by means of four different views that analyze the model from different perspectives:

- The *organizational view* that allows specifying the system from a high level of abstraction point of view. This view allows specifying the global purposes of the system, the relationships with its environment, the division of the functionality of the system in roles, and the main structure of the system.
- The *internal view* that allows specifying each entity (organizations, agents, and roles) of the system in high and low level of abstraction point of view. From a high level of abstraction, this view allows specifying the beliefs and objectives of each entity, and how the entity participate in the system and interact with its environment. From a low level of abstraction, this view allows specifying the internal functionality of each entity by means of the specification of which task and service implements. One instance of this view of the metamodel is created for each entity of the system.
- The *contractTemplate view* that allows specifying *contract templates* which are predefined restrictions that all final contract of a specific type must fulfill. Contracts are inherently defined at runtime, but contract templates are defined at design time and can be used at runtime as an initial point for the negotiation of contracts and to verify if the final contract is coherent with the legal context.
- The *activity view* that allows specifying interaction protocols, the sequence of activities in which a task or a service implementation is decomposed.

ROMAS metamodel is completely described in Chap. 5.

4.3 ROMAS Process Lifecycle

ROMAS tries to guide developers during the analysis and design phases in an intuitive and natural way. In that sense, ROMAS derives the whole design from the analysis of the requirements and their formalization by means of objectives. Following a goal-oriented approach, developers are focused from the early beginning in the purpose of the system.

ROMAS development process is composed of five phases, which help developers to analyze and design the system from the highest level of abstraction to the definition of individual entities and implementation details (Fig. 4.3). ROMAS phases are

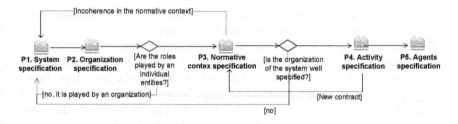

Fig. 4.3 The ROMAS process phases

completely detailed in Sect. 6.2. Following a summary of the purposes and results of each phase is presented:

- *Phase 1. System specification*: The purpose of this phase is to analyze the system requirements from a global point of view, i.e., focusing on the system as whole instead of focusing on the individual interests of each entity. These requirements are translated in terms of objectives and restrictions. The global objectives of the system are studied and refined into operational objectives, and the main use cases of the system are specified. Once all the requirements of the system have been analyzed, the last task of this phase is to evaluate the suitability of the ROMAS methodology for the development of the system regarding its specific requirements.

  The results of this phase of the methodology are: (1) a textual description of the system requirements, (2) a textual description of the objectives of the system, (3) an objective decomposition diagram, (4) a set of the use cases diagrams, and (5) a study of the suitability of the ROMAS methodology for this system.

- *Phase 2. Organization specification*: The purpose of this phase is to analyze and design the social structure of the system. First, the functionalities of the system are associated to roles. Then, the relationships between these roles, the restrictions, and the social environment of the system are analyzed in order to select the most suitable social architecture. This social architecture specifies in a high level of abstraction which are the social relationships between the roles of the system (like authority or collaboration) and if the system is composed of several organizations.

  The results of this phase of the methodology are: (1) a textual description of the roles of the system, (2) one diagram for each role of the system specifying its properties. These diagrams are instances of the Internal view of the metamodel, (3) one diagram for representing the social environment and structure of the system. This diagram is an instance of the organizational view of the metamodel.

- *Phase 3. Normative context specification*: The purpose of this phase is to formally specify the normative context of the system by means of norms and contracts. The requirements of the system, the normative documents associated to the system (like governmental legislation or institutional regulations), and the social structure of the system are analyzed in order to identify the norms and contracts that should be formalized. The processes of identification, formalization, and validation of the normative context are supported by a set of guidelines.

The results of this phase are: (1) modifications on the diagrams defined in the previous phase in order to add the norms and contracts identified, (2) a set of diagrams for specifying the contract templates of all the identified social relationships. These diagrams are instances of the contract template view of the metamodel.

- *Phase 4. Activity specification*: The purpose of this phase is to specify the tasks, services, and protocols that have been identified in the previous phases of the development process. In that sense, this phase revises the role internal view diagrams, the organizational view diagrams, and the contract template view diagrams in order to identify which tasks, services, and protocols should be detailed. For example, for each contract template, a negotiation and an execution protocol should be specified.

  The results of this phase are a set of diagrams, one for each task, service, and protocol, that are instances of the activity view of the metamodel.

- *Phase 5. Agents specification*: The purpose of this phase is to analyze and design every individual entity of the system. This phase analyzes the requirements of each entity, its restrictions, and which roles this entity should play in order to achieve its objectives. The last step of this phase is to validate the coherence between the design of every individual entity and the global design of the system.

  The results of this phase are a set of diagrams, one for each individual entity, that are instances of the internal view of the metamodel and that specifies the features, properties, and interactions of this entity with the rest of the system.

As shown in Fig. 4.3, this is not a linear process but an iterative one, in which the identification of a new element of functionality implies the revision of all the diagrams of the model and the work products produced, so it requires to go back to the appropriate phase. For example, during the second phase (Organization specification), part of the detected roles can be played by a group of agents that form another organization. In this case, it is necessary to go back to the first phase of the methodology to analyze the characteristics, global objectives, and structure of this organization.

4.4 ROMAS Background

The complete ROMAS background is presented in Chap. 2. The most remarkable influences for ROMAS are GORMAS [7] and OperA [40].

ROMAS uses the GORMAS metamodel as a starting point for the specification of its own metamodel. GORMAS is a service-oriented methodology that defines a set of activities for the analysis and design of organizational systems, including the design of the norms that restrict the behavior of the entities of the system. ROMAS metamodel inherits from GORMAS the concepts of agents, organizations, services,

Icon	Name	Description
	Phase	Phases represent significant periods in a project, ending with major management checkpoint, milestone, or set of deliverables. It is composed of a set of activities.
	Activity	Activities represent set of tasks. Activities are supposed to produce finer grained artifacts than phases.
	Task	Tasks represent actions that are performed during the development process. Tasks are supposed to concur to the definition of activity-level artifacts.
	In use	This icon is attached to any entity that is in use in this diagram.
	Role	Roles represent human entities that participate in a diagram.
	Guideline	Guidelines represent best practices suggested for a good application of the process documentation template or techniques about how to perform the prescribed work.
	Structured work product	It is a text document ruled by a particular template or grammar, for instance a table or a code document.
	Behavioral work product	It is a graphical kind of work product and is used to represent the dynamic aspect of the system (for instance a sequence diagram representing the flow of messages among agents along time);
	Structural work product	It is a graphical kind of work product and is used for representing the static aspect of the system, for instance a UML class diagram.
	Composite work product	It is a work product that can be made by composing the previous work product kinds, for instance a diagram with a portion of text used for its description.

Fig. 4.4 Summary of the SPEM 2.0 notation

and norms. ROMAS revises the GORMAS metamodel in order to refine these concepts. ROMAS also adds the concept of social and commercial contract. The process lifecycle of ROMAS and GORMAS are completely different. GORMAS bases the development process in the specification of the services that every entity must provide and use, while ROMAS bases it in the objectives of the system and the objectives of each entity of the system.

The graphical notation used in ROMAS to formalize the models is based on the notation used in GORMAS [7], ANEMONA [68], and INGENIAS [95]. ROMAS adds few graphical icons to represent some elements, like contract templates, that were not previously defined in these methodologies.

The concept of social contract used in ROMAS is similar to the concept of contract in the Opera methodology. However, ROMAS does not share the same concept of organizations and interactions. Organizations in OperA are defined as institutions where agents interact between them entering in previously determined scenes. Moreover, other differences are that OperA does not include the analysis and design of individual agents and it does not offer specific guidelines for identify the norms derived from the analysis of the requirements, legal documents, or design decisions.

4.5 FIPA Design Process Documentation Template

ROMAS methodology is described using the Design Process Documentation Template proposed by the FIPA Design Process Documentation and Fragmentation Working Group.[1] The complete specification of this standard can be consulted in [51].

This standard uses the SPEM 2.0 notation [103]. Figure 4.4 summarizes the most used concepts and shows their graphical icons in order to facilitate the understanding of the specification of the ROMAS methodology presented in the following sections.

The use of this standard template for specifying the ROMAS methodology is beneficial in the sense that:

- The use of the standard ensures that the whole development process is completely specified.
- The use of the standard facilitates the comparison with other methodologies described with the same standard.
- The use of the standard reduces the methodology learning time of developers used to this standard.
- This template is designed in order to facilitate the creation and reuse of method fragments. The template proposes the use of the Situational Method Engineering paradigm in order to provide means for constructing ad hoc software engineering processes following an approach based on the reuse of portions of existing design processes (method fragments). In that sense, our methodology could be extended by adding a method fragment from other methodology. Parts of our methodology could also be used to add functionality to other methodology or to create a new software engineering process for a specific purpose.

---

[1] http://www.pa.icar.cnr.it/cossentino/fipa-dpdf-wg/.

Chapter 5
ROMAS Modeling Language

This section details the ROMAS metamodel elements, relationships, and structure.

As is introduced above, ROMAS offers an unified metamodel that can be instantiated by means of four different views: Organizational view, Internal view, Contract Template view, and Activity view. A complete description of these views of the metamodel is presented in Sect. 5.1. Section 5.2 shows the graphical notation used to draw the ROMAS models.

Table 5.1 describes the entities that the ROMAS metamodel uses for modeling normative open MAS.

5.1 ROMAS Metamodel Views

5.1.1 Organizational View

In this view, the global goals of the organizations and the functionality that organizations provide and require from their environment are defined (Fig. 5.1). The static components of the organization, i.e., all elements that are independent of the final executing entities are defined too. More specifically, it defines:

- The entities of the system (*Executer*): *AAgents* and *Roles*. The classes *Executer* and *AAgents* are abstractions used to specified the metamodel, but neither of them are used by designers to model systems.
- An *AAgent* is an abstract entity that represents an atomic entity (*Agent*) or a group of members of the organization (*Organizational Unit*), seen as a unique entity from outside.
- The *Organizational Units* (OUs) of the system, that can also include other units in a recursive way, as well as single agents. The *Contains relationships* includes conditions for enabling a dynamical registration/deregistration of the elements of an OU through its lifetime.

© Springer International Publishing Switzerland 2015
E. Garcia et al., *Regulated Open Multi-Agent Systems (ROMAS)*,
DOI 10.1007/978-3-319-11572-6_5

Table 5.1 Definition of ROMAS metamodel elements

Concept	Definition	Metamodel views
Objective	An objective is a specific goal that agents or roles have to fulfill. It can be refined into other objectives	Organizational Internal view
Organizational Unit (OU)	A set of agents that carry out some specific and differentiated activities or tasks by following a predefined pattern of cooperation and communication. An OU is formed by different entities along its life cycle which can be both single agents or other organizational units, viewed as a single entity	Organizational Internal Contract template
Role	An entity representing part of the functionality of the system. Any entity that plays a role within an organization acquires a set of rights and duties	Organizational Internal Contract template Activity
Agent	An entity capable of perceiving and acting into an environment, communicating with other agents, providing and requesting services/resources and playing several roles	Organizational Internal Contract template Activity
Norm	A restriction on the behavior of one or more entities	Organizational Internal Contract template Activity
Contract template	A set of predefined features and restrictions that all final contract of a specific type must fulfill. A contract represent a set of rights and duties that are accepted by the parties	Organizational Internal Contract template Activity
Bulletin Board	A service publication point that offers the chance of registering and searching for services by their profile	Organizational Internal Contract template Activity
Product	An application or a resource	Organizational Internal Contract template Activity
Service profile	The description of a service that the agent might offer to other entities	Organizational Internal Activity
Service Implementation	A service specific functionality which describes a concrete implementation of a service profile	Internal Activity
Task	An entity that represents a basic functionality, that consumes resources and produces changes in the agent's Mental State	Organizational Internal Contract template Activity
Stakeholder	A group that the organization is oriented to and interacts with the OUs	Organizational

(continued)

Table 5.1 (continued)

Concept	Definition	Metamodel views
Belief	A claim that an agent (or a role taken by an agent) thinks that it is true or will happen	Internal
Fact	A claim that is true at the system's domain. The difference between beliefs and facts is the level of confidence in their veracity. While an agent is completely sure that its facts has happened, a belief is something that an agent hopes to happen or that thinks that has happened	Internal
Event	The result of an action that changes the state of the system when it occurs	Internal
Interaction	An entity defining an interaction between agents	Activity
Interaction unit	A performative employed during the interaction	Activity
Translation condition	An artifact that allows defining the sequence of tasks depending on a condition	Activity
Executer	A participant in an interaction. It can be an Organization, an Agent or a Role	Organizational Internal Contract template Activity

- The global *Objectives* of the main organization. The objectives defined in this view are nonfunctional requirements (softgoals) that are defined to describe the global behavior of the organization.
- The *Roles* defined inside the OUs. In the *contains* relationship, a minimum and maximum quantity of entities that can acquire this role can be specified. For each role, the *Accessibility* attribute indicates whether a role can be adopted by an entity on demand (external) or it is always predefined by design (internal). The *Visibility* attribute indicates whether entities can obtain information from this role on demand, from outside the organizational unit (public role) or from inside, once they are already members of this organizational unit (i.e., private role). A hierarchy of roles can also be defined with the *InheritanceOf* relationship.
- The organization social relationships (*RelSocialRelationship*). The type of a social relationship between two entities is related with their position in the structure of the organization (i.e., information, monitoring, supervision), but other types are also possible. Some social relationships can have a *ContractTemplate* associated which formalize some predefined commitments and rights that must be accepted or negotiated during the execution time. Each *Contract Template* is defined using the Contract Template view.
- The *Stakeholders* that interact with the organization by means of the publication of offers and demands of *Products* and *Services* in the *BulletinBoard*.

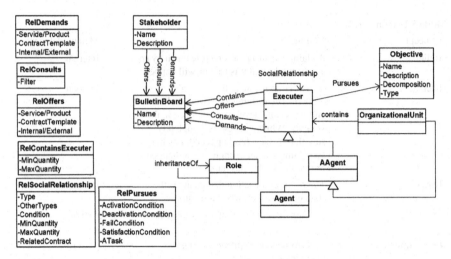

Fig. 5.1 Organizational view (The class *RelXXX* represents the attributes of the relationship *XXX*)

- The *Bulletin Board* can be considered as an information artifact for Open MAS. This artifact allows the designer to define the interaction with external entities and facilitates trading processes. When an agent wants to trade, the agent can consult or publish their offer into the BulletinBoard. Each offer or demand can be associated with a ContractTemplate. It means that this offer or demand has some predefined restrictions which are specified in this ContractTemplate view.

5.1.2 Internal View

This view allows defining the internal functionality, capabilities, beliefs, and objectives of each entity (organizations, agents, and roles) by means of different instances of this model (Fig. 5.2). More specifically, it defines the following features of each entity:

- The *Objectives* represent the operational goals, i.e., the specific goals that agents or roles have to fulfill. They can also be refined into more specific objectives. They might be related with a Task or Interaction needed for satisfying this objective.
- The *Mental States* of the agent, using beliefs, events, and facts.
- The *products* (resources/applications) available by an OU.
- The *tasks* that the agent is responsible for, i.e., the set of tasks that the agent is capable of carrying out. Task An entity that represents a basic functionality that consumes resources and produces changes in the agent's Mental State.

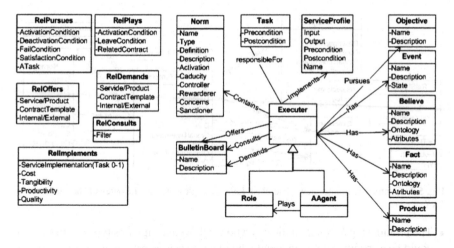

Fig. 5.2 Internal view (The class *RelXXX* represents the attributes of the relationship *XXX*)

- The *Implements Service Profile*.
- Internal entities can publish offers and demands in a *BulletinBoard*, as external stakeholder can do by means of the organizational view. This publications can also have an associated Contract Template to describe some predefined specifications.
- The *roles* that an agent or an organizational unit may play inside other organizational units (*Plays* relationship). *ActivationCondition* and *LeaveCondition* attributes of this relationship indicate in which situation an OU acquires or leaves a role.
- The *roles* played by each agent. *ActivationCondition* and *LeaveCondition* attributes of this *play* relationship indicate in which situation an agent can acquire or leave a role.
- The *Norms* specify restrictions on the behavior of the system entities. The relationship *Contains Norm* allows defining the rules of an organization and which norms are applied to each agent or role. *norms* that control the global behavior of the members of the OU.

5.1.3 Contract Template View

This view allows defining *Contract Templates*. Contracts are inherently defined at runtime. Despite this, designers represent some predefined restrictions that all final contract of a specific type should follow by means of a *contract template*. Contract templates can be used at runtime as an initial point for the negotiation of contracts and to verify if the final contract is coherent with the legal context. The syntax of a contract template is defined in Fig. 5.3. More specifically, it defines:

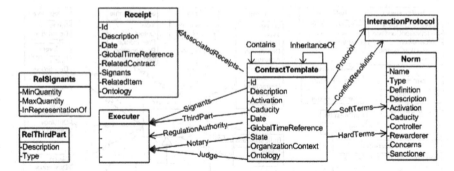

Fig. 5.3 Contract template view (The class *RelXXX* represents the attributes of the relationship *XXX*)

- The relationship *Signants* indicates who is allowed to sign this type of contract. It could be a specific agent, an agent who plays a specific role or an organization. A *ThirdPart* could be anyone who participates in the negotiation protocol or who is affected by the final execution of the Contract.
- The relationship *Protocol* indicates which protocols are recommended to negotiate this type of contract.
- After the negotiation, the *Notary* is responsible for verifying the correctness and coherence of the final contract definition. He should check if any term of a contract violate any norm of the regulated environment.
- Each type of contract can define which *Receipts* will be generated during the execution time. Receipts are proofs of facts, for example, a receipt can be generated when an agent successfully provides a service.
- In case of conflict, the *Judge* has to evaluate the Complaints and the generated Receipts following the *ConflictResolution* protocol. If he decides that there has been a violation of a norm, the *RegulationAuthority*, who is the main authority in the context of a contract, can punish or reward the agent behaviors.
- The relationship *Hard clause* indicates that any instance of this type of contract has to include this norm. *Soft clause* are recommendations, so during the negotiation stage *Signants* will decide whether this norm will be included or not in the final contract.

5.1.4 Activity View

This view allows defining the sequence of actions in which a task, a service, or a protocol can be decomposed (Fig. 5.4). Each *state* represents an action or a set of actions that must be executed. An *action* is a first-order formula that indicates which task or service is executed or which message is interchanged between the agents that participate in this state. The relationship *next* indicates the sequence of states. These sequences can be affected by a *translation condition* that indicates under which circumstances the state is going to be the next step of the process.

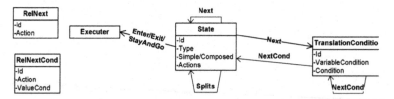

Fig. 5.4 Activity view (The class *RelXXX* represents the attributes of the relationship *XXX*)

5.2 ROMAS Notation

ROMAS models are graphically represented following the notation detailed in Fig. 5.5. This notation is based on the notation used in the GORMAS [6] and which was initially proposed by the INGENIAS methodology [95]. In order to represent the entities of the ROMAS metamodel that do not exist in these other methodologies like the abstraction of contract template, new graphical icons have been created.

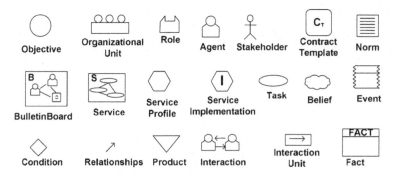

Fig. 5.5 Entities from the ROMAS graphical notation

Chapter 6
ROMAS Methodology

This chapter presents the ROMAS methodology which is a methodology for the analysis and design of normative open MAS. This chapter is organized as follows:

Section 6.1 presents a brief description of the case study that will be used during the whole chapter as a running example. Section 6.2 details the ROMAS process lifecycle by means of detailing its phases and activities. Section 6.3 presents the relationships between the work products produced and used during the process lifecycle as is specified in the FIPA standard. Finally, Sect. 6.4 summarizes the main conclusions and contributions of this chapter.

6.1 Case Study: Conference Management System

During the rest of the chapter we will use the Conference management system as a running case study in order to exemplify and clarify some parts of the development process. This case study deals with the development of a system to support the management of scientific conferences. This system involves several aspects from the main organization issues to paper submission and peer review, which are typically performed by a number of people distributed all over the world.

This system has been used previously as a case study by other methodologies such Tropos [90] or O-Mase [33]. In these works the normative context of the system and their entities are not studied. In this work, we consider that the Conference management system is regulated by a set of legal documents such the governmental law about data storage privacy, and that each conference managed in the system can define its own internal regulations.

6.2 Phases of the ROMAS Process

In this section, the phases that compose the ROMAS methodology are described following the FIPA standard *Design Process Documentation Template*. The description of each methodology is composed of the following parts:

© Springer International Publishing Switzerland 2015
E. Garcia et al., *Regulated Open Multi-Agent Systems (ROMAS)*,
DOI 10.1007/978-3-319-11572-6_6

- A brief introduction that summarizes the purposes of this phase. This introduction also includes two diagrams, one for representing the flow of activities of this phase and another for representing the relationships between the activities, tasks, roles and work products.
- *Process roles* subsection that lists the roles involved in the work of this phase and clarifies their level of involvement in the job to be done.
- *Activity details* subsection that describes the sequence of tasks that are performed in each activity. This subsection presents a table where every task is detailed. Following the description of these tasks, developers can know exactly the sequence of actions that should be performed and which guidelines support them.
- *Work products* subsection that presents a the work products used and produced at each phase summarizing them in a table. This subsection also describes the relationship between the work product and the ROMAS metamodel and details the structure and the associated guidelines to every work product.

6.2.1 PHASE 1: System Specification

During this phase the analysis of the system requirements, global goals of the system and the identification of use cases are carried out. Besides, the global goals of the organization are refined into more specific goals, which represent both functional and non-functional requirements that should be achieved. Finally, the suitability of the ROMAS methodology for the specific system to be developed is analyzed.

The process flow at the level of activities is reported in Fig. 6.1. The process flow inside each activity is detailed in the following subsections (after the description of process roles). Figure 6.2 describes the System specification phase in terms of which roles are involved, how each activity is composed of tasks, which work products are produced and used for each task and which guidelines are used for each task.

6.2.1.1 Process Roles

There are two roles involved in this phase: the system analyst and the domain expert. The *domain expert* is responsible for: (1) describing the system requirements, by means of identifying the system main objectives, the stakeholders, the environment of the organization and its restrictions; (2) supporting the system analyst in the analysis of the objectives of the system; (3) supporting the system analyst in the description of the use cases of the system. The *system analyst* is responsible for: (1) analyzing the objectives of the system; (2) identifying the use cases; and (3) evaluating the

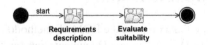

Fig. 6.1 The system description phase flow of activities

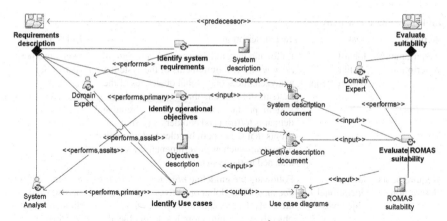

Fig. 6.2 The system description phase described in terms of activities and work products

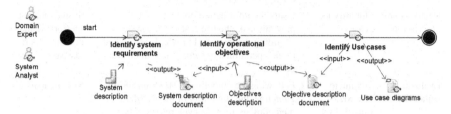

Fig. 6.3 The flow of tasks of the requirements description activity

suitability of the ROMAS methodology for the system to be developed regarding its requirements.

6.2.1.2 Activity Details

As Fig. 6.1 shows, this phase is composed of two activities: the *Requirements description* whose process flow is detailed in Fig. 6.3, and the *Evaluate suitability* that is composed of only one task. The first activity analyzes the requirements of the system to be developed. The second activity evaluates the suitability of the ROMAS methodology for analyzing and designing a system with these requirements.

All the tasks of this phase are detailed in Table 6.1. The description of the tasks details the sequence of actions that should be performed in this phase, which guidelines and work products are used by each task, and which work products are produced.

6.2.1.3 Work Products

This section details the work products produced in this phase. A brief description of these work products is presented in Table 6.2. Following each work product is detailed and their use is exemplified by means of our running example.

Table 6.1 Phase 1: activity tasks

Activity	Task	Task description	Roles involved
Requirements description	Identify system requirements	Following the guideline *system description document*, the requirements of the system are analyzed, including global objectives of the system, stakeholders that interact with the system, products and services are offered and demands to/from stakeholders, external events that the system handles and normative documents such as governmental laws attached to the system	Domain expert (performs)
Requirements description	Identify operational objectives	Following the guideline *objective description document*, the global objectives of the system are analyzed and split into operational objectives, i.e., into more low level objectives that can be achieved by means of the execution of a task or a protocol	System analyst (assists) and domain expert (performs)
Requirements description	Identify use cases	Using the information obtained in the previous task, the use cases of the system regarding the tasks and protocols associated to the operational objectives identified are defined	System analyst (performs) and domain expert (assists)
Evaluate suitability	Evaluate ROMAS suitability	Following the guideline *ROMAS suitability guideline*, the suitability of the ROMAS methodology for the development of the system to be developed regarding its specific features is evaluated	System analyst

Table 6.2 Phase 1: work products

Name	Description	Work product kind
System definition	This document analyzes the main features of the system and the relationship with its environment	Structured text
Objectives description	This document analyzes the global objectives of the system and decomposes them into operational objectives	A composite document composed by a structured text document and a diagram
Use cases	These diagrams are UML graphical representations of workflows of stepwise activities and actions with support for choice, iteration and concurrency	Behavioral
ROMAS suitability guideline	It is a questionnaire to evaluate the suitability of the ROMAS methodology for the development of the analyzed system	Structured text

System description document

This document is employed to identify the main features of the system and its relationship with the environment. Table 6.3 shows the template that describes each analyzed system attribute.

Table 6.3 Template for system description document

Property	Description		Guideline
System identifier	General name of the system to be developed		It is recommended to select a short name or an abbreviation
System description	Informal description of the system		There is no limitation on the length of this text. • What is the motivation for developing such a system? • Is there any system requirement that specifies if the system must be centralized or decentralized? • Which is the main objective of this system?
Domain	Domain or domains of application		If this system must be able to be applied in different domains, it is recommended to add a text that explains each domain and whether it is necessary to adapt the system to each domain
Kind of environment	Identify and specify the kind of environment of the system		• Can the functionality of the system be distributed between different entities? • Are the resources of the system distributed in different locations? • Are there external events that affect the internal state and behavior of the system? Is it a reactive system? • Is it a physical or a virtual environment? Is there any physical agent or robot that plays a role in the system? • Is there any human interaction with the system? • Should the results of the system be presented graphically? Is there any graphical environment?
Global objectives	Functional and non-functional requirements (softgoals) that specify the desired-global behavior of the system		• Which are the purposes of the system? • Which results should provide the system? • Should the system keep any parameter of the system between a specific threshold? (ex. the temperature of the room, the quantity of money in an account and so on)
Stakeholders	Identifier	An identifier for the stakeholder	Are there external entities or applications that are able to interact with the system?
	Description	Informal description of the stakeholder	
	Type	Indicate if the stakeholder is a client, a provider or a regulator	

(continued)

Table 6.3 (continued)

Property	Description		Guideline
	Contribution	To point out what the organization obtains from its relationship with the stakeholder	
	Requires	A set of products and/or services that the stakeholder consumes	
	Provides	A set of products and/or services that the stakeholder offers to the organization	
	Frequency	To point out whether this stakeholder contacts with the organization frequently, occasionally or in an established period of time	
Resources	Resources and applications available by the system		• Is there any application or resource available by the system? • Is this resource physical or virtual?
Events	External events that produce a system response		Which events can produce an effect on the system? How the system capture these events and how response to them?
Offers	A set of products or services offered by the organization to its clients		Is there any product or service that the system should provide to an external or internal stakeholder?
Demands	A set of products or services demanded by the organization to its clients		Are there any requirements that the system cannot provide itself? Is it important who provide this service or product?
Restrictions	An overview about which types of restrictions the system should imposed on its entities		• Behavioral restrictions: Is there any system requirement that specifies limits on the behavior of the members of the system? • Critical restrictions: Is there any action whose inadequate usage could be dangerous for the system? • Usage restrictions: Is there any restriction on the usage of the system resources? Is there any restriction on the usage of the services and products offered by the system? Is there any restriction on who is an appropriate stakeholder to provide a service or product to the system? • Legal restrictions: Is there any normative document, such as governmental law or institutional internal regulations, that affects the system's entities behavior?

Table 6.4 Phase 1—case study: system description document

Case study: *System description document*	
System identifier	Conference Management System (CMS)
System description	This system should support the management of scientific conferences. This system involves several aspects from the main organization issues to paper submission and peer review, which are typically performed by a number of people distributed all over the world
Domain	Research
Kind of environment	Virtual and distributed environment with established policies and norms that should be followed
Global objectives	• Management of user registration
	• Management of conference registration
	• Management of the submission process
	• Management of the review process
	• Management of the publication process
Stakeholders	There is no external entity that interact with the system. Every entity that wants to interact with the system should be registered and logged in the system
Resources	Database: it should include personal information and affiliation and information about which users are registered as authors, reviewers or publishers for each conference. Also it should include information about each conference, i.e., its status, its submitted papers and reviews, ...
Events	Non external events are handled by the system
Offers	• NewUsers_registration()
	• Log_in()
Demands	
Restrictions	• The system should follow the legal documentation about the storage of personal data
	• Each conference should describe its internal normative

Table 6.4 shows how this template has been used to analyzed the CMS case study. This document shows that the CMS system is a distributed system in a regulated environment with a set of global objectives. The system should offer to external entities two services, one for registering in the system and another to log in.

Objectives description document

This document analyzes the global objectives of the system and decomposes them into operational objectives. It is a composite document composed by a structured text document and a diagram.

The structured text document template is shown in Table 6.5. Every global objective specified in the *system description document* is described using this document. The global objectives of the systems are refined into more specific ones that should also be described using this document. The document will be completed when all the global objectives are decomposed into operational objectives, i.e. they are associated

Table 6.5 Phase 1: objectives description

Property	Description	Guideline
Identifier	Objective name identifier	It is recommended to select a short name or an abbreviation
Description	Informal description of the objective that is pursued	There is no length limitation on this text. It should clearly describe this objective
Activation condition	First order formula that indicates under which circumstances this objective begins being pursued	• Does the organization pursue this objective from the initialization of the system? • Is there any situation that activates this objective? Common circumstances that can activate objectives are: when an event is captured, the failure of other objective, the violation of a restriction, when an agent plays a specific role, and so on • If this objective is deactivated, is there any situation that forces the objective to be pursued again?
Deactivation condition	First order formula that indicates under which circumstances this objective stops being pursued	• Is this objective pursued during the whole lifecycle of the system? • Is there any situation that deactivates this objective? Common circumstances that deactivate an objective are: when it is satisfied, when other objective is satisfied, when some restriction has been violated, and so on
Satisfaction Condition	First order formula that indicates in which situation this objective is satisfied	• Is the satisfaction of this objective measurable? • What results should be produced to claim that this objective has been satisfied?
Fail Condition	First order formula that indicates in which situation this objective has failed	• Is there any situation that is contrary to this objective and that will invalidate it? • Is there any threshold that should not be exceeded?
Type	Objectives can be abstract or operational. An *abstract objective* is a non-functional requirement that could be defined to describe the global behavior of the organization. An *operational objective* is a specific goal that agents or roles have to fulfill	If there is a task that can be executed in order to satisfy this objective, it is an operational objective, in other cases it is an abstract objective. Abstract objectives can be refined into other abstracts or operational objectives
Decomposition	First order formula that specified how this objective is decomposed	If this is an abstract objective it should be decomposed in several operational objectives which indicates which tasks should be executed in order to achieve this objective. Operational objectives can also be decomposed in order to obtain different subobjectives that can be pursued by different members of the organization. This fact simplifies the programming task and facilitates the distribution of responsibilities

(continued)

Table 6.5 (continued)

Property	Description	Guideline
Related Action/ Restriction	Objectives can be related to a restriction on the behavior of the system, or to an action that must be executed in order to achieve this objective. Actions can be tasks, services or protocols	The difference between a task and a protocol is that a task can be executed by one single agent, however a protocol is a set of tasks and interactions between two or more agents. Services are pieces of functionality that an entity of the system offers to the others, so the main difference between services and tasks or protocols is that they are executed when an entity request this functionality. At this phase it is not necessary to detail all the parameters of the task. You should describe in a high abstraction level what actions and activities are necessary to achieve this objective
	Type	Task, service or protocol	
	Identifier	An identifier for the task, service or protocol	
	Description	Informal description of the action	
	Resources	Which applications or products are necessary to execute this task (for example, access to a database). This feature can be known at this analysis phase due to requirement specifications. However, if there is no specification, the specific implementation of each task should be defined in following steps of the methodology	
	Activation condition	First order formula that indicates under which circumstances this action will be activate	
	Inputs	Information that must be supplied	
	Precondition	A set of the input conditions and environment values that must occur before executing the action in order to perform a correct execution	
	Outputs	Information returned by this action and tangible results obtained	
	Post-conditions	Final states of the parameters of the environment, by means of the different kinds of outputs	

to tasks, protocols or restrictions that must be fulfilled in order to achieve these objectives. It is recommended to create one table for each global objective. The first column of each table will contain the properties name, the second the description of the global objective and the following columns the descriptions of the objectives in which this global objective has been decomposed. As an example of the decomposition of a global objective into operational ones, Tables 6.6 and 6.7 show the decomposition of the global objective *Conference registration*. The abstract objective of *Conference registration* is decomposed in two objectives: *Create new conference* and *Allow supervision*. In the same way the objective *Allow supervision* is decomposed in three operational objectives: *Modify conference details*, *Get information about submission*, *Get information about reviews* and *Validate reviews decision*. The details about these objectives are presented in these tables.

The diagram represents graphically the decomposition of the objectives by means of an UML diagram in order to provide a general overview of the purpose of the system that can be easily understood by domain experts. The graphical overview of the CMS case study objectives is shown in Fig. 6.4, where A means abstract objective and O means operational objective.

Use case diagram

These diagrams are UML graphical representations of workflows of stepwise activities and actions with support for choice, iteration and concurrency. The actions identified in the analysis of the operational objectives are related forming activity diagrams in order to clarify the sequence of actions that will be performed in the system. The idea is not to completely describe each task and neither detail who is the responsible of it. With these diagrams what the System Analyst should clarify is the sequence of actions and the possibility of choice, iteration and/or concurrency. Figure 6.5 shows the sequence of actions that can be performed in the CMS case study. It shows that to have access to the functionality of the system, users need to log in first. It also shows that there are a set of activities that should be performed sequentially, for example, after registering a new submission authors should be informed of the status of their submission.

ROMAS suitability guideline

After analyzing the requirements of the system, it is recommended to use this guideline in order to evaluate the suitability of the ROMAS methodology for the development of the analyzed system. Table 6.8 shows the criteria used to evaluate whether ROMAS is suitable.

ROMAS is focused on the development of regulated multi-agent systems based on contracts. ROMAS is appropriate for the development of distributed system, with autonomous entities, with a social structure, with the need of interoperability standards, regulation and trustworthiness between entities and organizations. ROMAS is not suitable for the development of centralized systems or non multi-agent systems. Although non normative systems could be analyzed using ROMAS, it is not recommended.

Table 6.6 Phase 1—case study: objective description document I

Case study: *Conference registration objective* decomposition I

Identifier	Create new conference	Allow supervision	Modify conf details
Description	The system should allow the registration of new conferences. The entity who registers the conference should be the chair of it	The authorized entities should be able to supervise the status of the conference and modify its details	The description details such as deadlines, topics and general description can be modified by the chair or vice-chair of the conference
Activation condition	True (always activated)	Conference_status= activated	Conference_status= activated
Deactivation condition	False	Conference_status= cancelled	Conference_status= cancelled
Satisfaction condition			
Fail Condition			
Type	Operational	Abstract	Operational
Decomposition		Modify conf details AND Get info submissions AND Get info reviews AND Validate reviews decision	

Related Action/Restriction

Identifier	Create_new_conference()		Modify_conf_details()
Type	Service		Service
Description	The registration must be performed by means of a graphical online application		After checking that the user that is trying to modify the conference details is authorized to do that, the system will provide a graphical online application to update the details. The information is shown by means of a graphical online application
Resources	Access to the conferences database		Access to the conferences database
Activation condition	Registered user demand		
Inputs	Deadlines, topics of interests and general information		
Precondition	The entity that executes the task should be a registered user		
Outputs			
Postconditions	The user that executes the task becomes the chair of the conference		

Table 6.7 Phase 1—case study: objective description document II

Case study: *Conference registration objective* decomposition II

Identifier	Get Info submissions	Get Info reviews	Validate reviews decision
Description	The system should provide information about the submitted papers	The system should provide information about the reviews that has been uploaded in the system	The chair should validate the decisions about acceptance or rejection of papers performed by the reviewers
Activation condition	Conference_status= activated	Conference_status= activated	Conference_status= revision
Deactivation condition	Conference_status= cancelled	Conference_status= cancelled	Conference_status= cancelled
Satisfaction condition			
Fail condition			
Type	Operational	Operational	Operational
Decomposition			
Related Action/Restriction			
Identifier	Get_Info_ submissions()	Get_Info_ reviews()	Validate_reviews_ decision()
Type	Service	Service	Task/Protocol
Description	Only pc members can access to the information about submissions. The information is shown by means of a graphical online application	Only pc members that are not authors of the paper can access to the reviews of a specific paper. The information is shown by means of a graphical online application	The chair should validate one per one the decision for each paper. If the chair performs the action by itself this objective would be pursued by means of a task. If the final decision is negotiated between the PC members this objective should be pursued by means of a protocol
Resources	Access to submitted papers database	Access to reviews database	Access to reviews database
Activation condition			After the review deadline is finished
Inputs			
Precondition			
Outputs			
Postconditions			After the validation the decision is considered final and authors should be notified

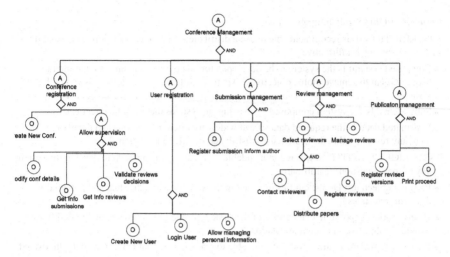

Fig. 6.4 Case study: objective decomposition diagram

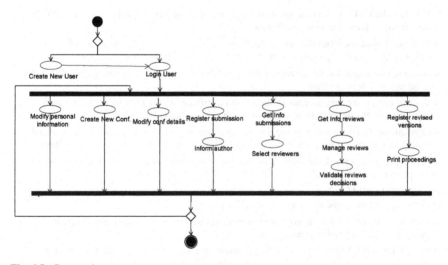

Fig. 6.5 Case study: use case

The analysis of the CMS case study features following this guideline shows that ROMAS is suitable for the development of this system. It is a distributed system, composed by intelligent systems with social relationships between them. The behavior of the system's entities and their relationships are bounded by the regulations of the system. The rights and duties that an entity acquires when it participates in the system should be formalized. For example, reviewers should know before acquiring the commitment of reviewing a paper, when its revision must be provided. Therefore, a contract-based approach is recommendable.

Table 6.8 ROMAS suitability guideline

DISTRIBUTION: It is recommendable to use a distributed approach to develop the system if any of these questions is affirmative

- Composed system: Is the system to be developed formed by different entities that interact between them to achieve global objectives? Are there different institutions involved in the system?
- Subsystems: Is the system composed by existing subsystems that must be integrated?
- Distributed data: Is the required data spread widely in different locations and databases? Are there any resources that the system uses distributed in different locations?

INTELLIGENT ENTITIES: It is recommendable to use an agent approach to develop the system if any of these questions is affirmative

- Personal objectives: Do the entities involved in the system have individual and potentially different objectives?
- Heterogenous: Is possible that entities of the same type had been implemented with different individual objectives and implementations?
- Proactivity: Are the entities of the system able to react to events and also able to act motivated only by their own objectives?
- Adaptability: Should be the system able to handle dynamic changes in its requirements and conditions?

SOCIAL STRUCTURE: It is recommendable to use an organizational approach to develop the system if any of these questions is affirmative

- Systems of systems: Does the system needs the interaction of existing institutions between which exist a social relationship in the real-world that must be taken into account?
- Social relationships: Do the entities of the system have social relationships, such as hierarchy or submission, between them?
- Departments: Is the functionality of the system distributed in departments with their own objectives but that interact between them to achieve common objectives?
- Regulations: Are there different regulations for different parts of the system, i.e. is there any regulation that should be applied to a group of different entities but not to the rest of them?
- Domain-like concepts: Is the domain of the system in the real-world structured by means of independent organizations?

INTEROPERABILITY: The system must implement interoperable mechanism to comunicate entities if any of these questions answers is affirmative

- Technical Interoperability: Is possible that different entities of the system use different (potentially incompatible) technologies?
- Process Interoperability: Is possible that different entities of the system employ divergent (potentially incompatible) processes to achieve their goals?
- Semantic Interoperability: Is possible that different entities of the system utilise different vocabularies and coding schemes, making it difficult to understand the data of others?

REGULATIONS: If the system has regulations associated it is recommended to apply a normative approach to develop the system. Only in the unlikely possibility that the norms of the system were static (no possibility of changing over time) and all the entities of the system are implemented by a trustworthy institution taking into account the restrictions of the system a non normative approach could be used

- Normative documents: Is the system or part of it under any law or institutional regulation?
- Resources restrictions: Are there specific regulations about who or how system resources can be accessed?

(continued)

Table 6.8   (continued)

- Dynamic regulations: Should the system be adapted to changes in the regulations?
- Openness: Is the system open to external entities that interact and participate in the system and these entities should follow the regulations of the system?
- Risky activities: Is there any action that if it is performed the stability of the system would be in danger?

TRUSTWORTHINESS: It is recommended to use a contract-based approach if any of these questions is affirmative

- Formal interactions: Are there entities that depend on the behavior of the others to achieve their objectives and whose interactions terms should be formalized?
- Contractual commitments: Should the entities of the system be able to negotiate terms of the interchanges of products and services and formalize the results of these negotiations?
- Social commitments: Are the entities of the system able to negotiate their rights and duties when they acquire a specific role? Could the social relationships between agents be negotiated between them?
- Control system: Is the system responsible of controlling the effective interchange of products between entities?
- Openness: Is the system open to external entities that interact and participate in the system acquiring a set of rights and duties?

6.2.2 PHASE 2: Organization Specification

During this phase the analysis of the structure of the organization is carried out. In the previous phase of the methodology, the operational objectives are associated to specific actions or restrictions. In this phase, these actions and restrictions are analyzed in order to identify the roles of the system. A *role* represents part of the functionality of the system and the relationships between roles specify the structure of the system.

The process flow at the level of activities is reported in Fig. 6.6. The process flow inside each activity is detailed in the following subsections (after the description of process roles). Figure 6.7 describes the Organization specification phase in terms of which roles are involved, how each activity is composed of tasks, which work products are produced and used for each task and which guidelines are used for each task.

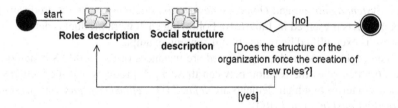

Fig. 6.6   The organization description phase flow of activities

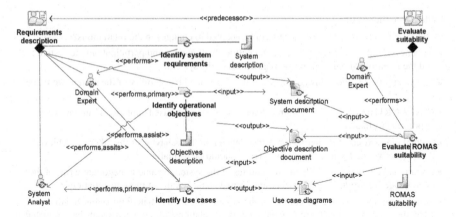

Fig. 6.7 The organization description phase described in terms of activities and work products

6.2.2.1 Process Roles

The roles involved in this phase are the same than in the previous phase: the *system analyst* and the *domain expert*. The domain expert is in charge of supporting the system analyst facilitating information about domain requirements and restrictions.

6.2.2.2 Activity Details

As Fig. 6.6 shows this phase is composed of two activities: *Roles description* and *Social structure description*. Each activity is composed of several tasks that are executed sequentially. All the tasks of the phase are detailed in Table 6.9. The description of the tasks details the sequence of actions that should be performed in this phase, which guidelines and work products are used by each task, and which work products are produced.

6.2.2.3 Work Products

This phase uses the work products produced in the previous phase (*Use cases* diagram, *System definition* and *Objective description* documents), and it produces the work products presented in Table 6.10. Following each work product is detailed and their use is exemplified by means of our running example.

Some of the work products generated are instances of the ROMAS metamodel. Figure 6.8 describes the relation between these work products and the metamodel elements in terms of which elements are *defined* (D), *refined* (F), *quoted* (Q), *related* (R) or *relationship quoted* (RQ).

Table 6.9 Phase 2: activity tasks

Activity	Task	Task description	Roles involved
Roles description	Identify roles	Following the guideline *Role identification guideline* the roles of the system are identified and associated to different parts of the system functionality	System analyst (assists) and domain expert (performs)
Roles description	Describe roles	Following the guideline *Role description document* each identified role is analyzed	System analyst (assists) and domain expert (performs)
Roles description	Represent roles	The details about each role are graphically represented by means of instances of the *internal view diagram*	System analyst (performs)
Social structure description	Identify organizational structure	Identify how the members of the organization interact between them, i.e., which social structure has the organization	System analyst (performs) and domain expert (assist)
Social structure description	Represent social structure	Represent graphically the identified social structure using an *organizational view diagram*	System analyst (performs)

Table 6.10 Phase 2: work products

Name	Description	Work product kind
Role identification guideline	This guideline supports the identification of the roles of the system	Structured text
Role description document	This document analyzes the main features of each role. It describes each role's objectives, resources and restrictions	Structured text
Internal view diagram	One internal view diagram is created for each role. They graphically represent the specification of each role	Behavioral
Organizational view diagram	This diagram represents graphically the structure of the system, its global objectives and the relationship with its environment	Behavioral

Role identification guideline

A role is an entity representing a set of goals and obligations, defining the services that an agent or an organization could provide and consume. The set of roles represents the functionality of the system, therefore the roles that a system should have are defined by the objectives of the system and should also take into account previous system requirements. The relationships and interactions between roles are the basis to define the structure of the organization. This guideline is designed to help the System Analyst to identify the roles that are necessary in the system. Figure 6.9 represents de sequence of activities to do.

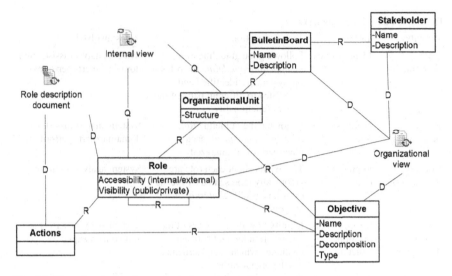

Fig. 6.8 Phase 2: relations between work products and metamodel elements

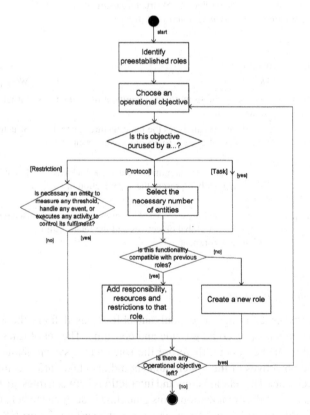

Fig. 6.9 Phase 2: role identification guideline

The first step of the process consists in asking the domain expert and check in the *system description* document whether there is any preestablished role defined in the requirements of the system.

 After that every operational objective described in the *Objective description document* should be analyzed. It is recommended to analyze all the operational objectives obtained by the decomposition of an abstract objective before analyzing the next abstract objective.

If this operational objective is associated to a restriction, it would add a norm in the organization that pursues these objectives. Besides, if this restriction is associated to a external event or a threshold there must be an entity responsible of handling this event or measuring this threshold variable.

If this operational objective is associated to a protocol, the system analyst should revise the sequence of actions necessary to perform this protocol in order to obtain all the entities that participates in this protocol.

Usually each task and functionality is associated to a role in order to create a flexible and distributed system. However, decomposing the system in too many entities can increase the number of messages between entities, the number of restrictions, and the complexity of each activity. Although, the System Analyst is the responsible of finding the balance taking into account the specific features of the domain, here we present some general guidelines:

It is not recommended to assign two functionalities to the same role when:

- These functionalities have different physical restrictions, i.e., they must be performed in different places.
- These functionalities have temporary incompatible restrictions, i.e., when they cannot be executed at the same time by the same agent. For example, it is usual that an entity was able to buy and sell, but as far as he is not able to sell and buy the same item at the same time, it is recommended to create one role Buyer and one role Seller. Remember that roles represent functionality, so any final entity of the system could be able to play several roles at the same time.
- These functionalities involve the management of resources that are incompatible. For example, the functionality of validating who is able to access to a database should not be join to the functionality of accessing to the database. The reason is that if the entity who is accessing to the database is the responsible of validating its own access, there can be security issues.

It is recommended to assign two functionalities to the same role when:

- These functionalities cannot be executed concurrently and they are part of a sequence.
- These functionalities access to the same resources and have the same requirements.
- These functionalities can be considered together as a general functionality.

In order to provide a fast and general overview, it is recommended to create a graphical representation of the relationships between the identified roles and the tasks and protocols. A relationship between a role and an action in this diagram means that the role is responsible from this action, participates in it or it is affected

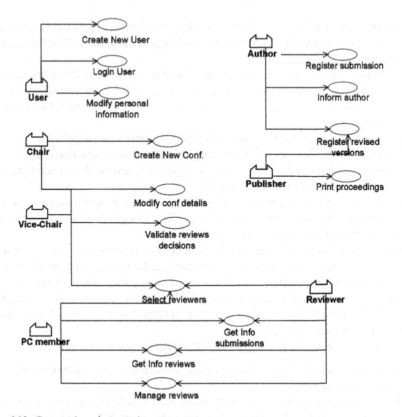

Fig. 6.10 Case study: roles overview

by its results. Figure 6.10 gives an overview of the results obtained when applying this guideline to the CMS case study. As is shown, seven roles has been identified:

- The *User* role is an entity of the system that must be registered in order to access to the system. On the contrary of the rest of the roles, this role is not related to any specific conference.
- The *Author* role is an entity attached to a specific conference in which this role can submit papers and receive information about the status of its papers.
- The *Chair* role is the main responsible from a conference. This role is able to create a conference and share the responsibility from selecting the reviewers, validate the revisions and update the conference details with the *Vice-Chair* role.
- The *PC member* role is responsible from managing the reviews, can participate in the selection of the reviewers and have access to the information about submissions and reviews for a specific conference.
- The *Reviewer* role is responsible from submit the reviews to the system.
- The *Publisher* role is responsible from managing the revised versions of the papers and print the proceedings.

Role description document and internal view diagram

Each role should be described by means of the guideline offered in Table 6.11. This guideline allows the analysis of the roles and also the analysis of the relationships between them. After this analysis, this information is graphically represented by means of an *internal view diagram* for each role.

Table 6.11 Phase 2: role description document

Property	Description	
Identifier	General name of role. It is recommended to select a short name or an abbreviation	
Description	Informal description of the role. There is no length limitation on this text	
List of objectives in which the role participates	Objective's identifier	The identifier of the objective that this role is going to contribute to its satisfaction
	Description of its contribution	An informal text describing how this role contributes to the satisfaction of this objective
	Task/ Protocol/ Service	Which task is responsibility of this role or in which protocol this role participates. If this task should be activated as a reaction of a petition of other entity, this task should be published as a service
	Responsibility shared with	A text explaining if this role pursue this objective alone or if he collaborates with others to achieve it
Resources: Used	A list of the resources (products, services and applications) that this role requires to develop its functionality. This text should specify which type of access the role needs (reading, executing, writing, partial or full access)	
Resources: Provided	A list of the resources (products, services and applications) that this role provides	
Events	A list of the events that this role handlesa and with which task	
Restrictions	A list of the restrictions that are inherent to the functionality that this role executes. These restrictions are mainly derived from the information provicesd by the Domain Expert	
Other memberships	A text explaining if this role in order to executes a task inside the organization it must be part of other different organization. If it is know, its rights and duties in the other organization must be detailed in order to ensure the coherence its objectives, rights and duties. However, due to privacy restrictions it is probable that many details cannot be shared between organizations	
Personal objectives	A role can pursue an objective not directly related to the functionality required by the organization. For example, it can pursue an objective in order to maintain its integrity	
Who plays the role	Is this role played by a single entity or by an organization? If it is played by an organization this organization must be analyzed following each step of the methodology	

As an example, Table 6.12 shows the description of the role *reviewer* from the case study. Figure 6.11 shows its graphical representation using a ROMAS internal view diagram.

Organizational view diagram

One organizational view diagram is created to graphically represent the structure of the system. Besides, this diagram also describes the overview of the system by means of its global *objectives* and how the system interact with its environment of the system (which *services* offers and demands to/from the *stakeholders* and which *events* the system is able to handle). The necessary information to fulfill these diagram is obtained from the *System description document*. Due to the fact that in the literature there are several well-defined guidelines to identify the organizational structure of a system, ROMAS does not offer any new guideline. Instead the use of the guideline defined by the GORMAS methodology in [9] is recommended.

Figure 6.12 shows the organizational view diagram of the CMS system case study. Inside the main system, the *Conference organization* represents each conference that is manage by the system. Each conference is represented as an organization because using this abstraction each one can define its own internal legislation and can refine the functionality assigned to each entity of the system.

6.2.3 PHASE 3: Normative Context Specification

During this phase the normative context of the system is analyzed by means of identifying and formalizing the norms and the social contracts that regulate the entities' behavior inside the system. As is described in the metamodel (Chap. 5), norms are formalized using the following syntax:

   (normID, Deontic, Target, Activation, Expiration, Action, Sanction, Reward).

The process flow at the level of activities is reported in Fig. 6.13. The process flow inside each activity is detailed in the following subsections (after the description of process roles). Figure 6.14 describes the Normative context specification phase in terms of which roles are involved, how each activity is composed of tasks, which work products are produced and used for each task and which guidelines are used for each task.

6.2.3.1 Process Roles

The system analyst is responsible for performing the activities of this phase. The domain expert will support the system analyst during the identification of the norms that regulate the system.

Table 6.12 Phase 2: case study—reviewer role description document

Property	Description		
Identifier	Reviewer		
Description	This role is responsible from submit the reviews to the system. It is attached to a specific conference. It is responsible from submit a review from a paper within the established deadline and in the specific format that the conference specifies		
List of objectives in which the role participates	Objective's identifier	Select reviewers	Manage reviews
	Description of its contribution	Reviewers should negotiate with *PC members* which papers are they going to review, when they are supposed to provide the reviews and which specific format these reviews must have	Reviewers should send to the system their reviews. PC members would validate the reviews and contact the reviewers if there is any doubt in the information supplied
	Task/Protocol/ Service	Reviewers participate in the protocol Select_reviewers()	Reviewers are responsible from the protocol *Manage_reviews()* and they offer the service *Execute_review()*
	Responsibility shared with	PC memebers	
Resources: Used	• Reviews and papers database • They use the service *Get_info_submissions()*		
Resources: Provided			
Events	Conference details modification event		
Restrictions	• The same entity cannot be the author and the reviewer of the same paper • Reviewers only have access to the information about their own reviews. They do not have access to other reviews or to the authors' personal details		
Other memberships	Any entity that wants to play the role reviewer should be previously registered in the system as a *user*		
Personal objectives	In general, there is no personal objectives for reviewers in the system. However, some conferences can encourage the productivity of their reviewers by offering rewards for each revised paper or for presenting the reviews before a specific date		
Who plays the role	This role is played by a single entity		

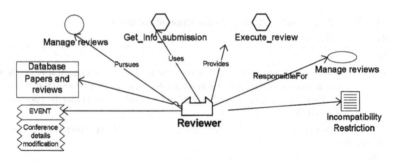

Fig. 6.11 Case study: reviewer role diagram

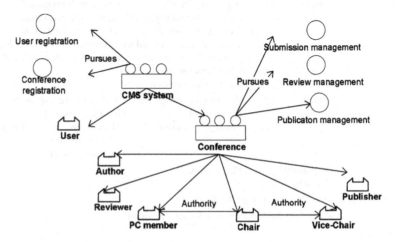

Fig. 6.12 Case study: organizational diagram

Fig. 6.13 Phase 3: activity tasks

6.2.3.2 Activity Details

As Fig. 6.13 shows this phase is composed of tree activities: *Identify restrictions from requirements*, *Identify social contracts* and *Verify normative context*.

Each activity is composed of one task. All the tasks of this phase are detailed in Table 6.13. The description of the tasks details the sequence of actions that should be performed in this phase, which guidelines and work products are used by each task, and which work products are produced.

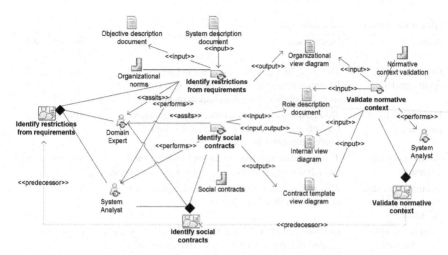

Fig. 6.14 Phase 3: resources and products used

Table 6.13 Phase 3: activity tasks

Activity	Task	Task description	Roles involved
Identify restrictions from requirements	Identify restrictions from requirements	Following the guideline *Organizational norms*, the system analyst formalizes the norms described in the requirements that regulate the agent behavior. This norms refine the *organizational view* diagram of the organization associated to these norms Following the guideline *Normative documents*, the system analyst extracts from the normative documents attached to the system requirements the norms and restrictions that must be integrated in the design	System analyst and domain expert
Identify social contracts	Identify social contracts	Following the guideline *Social contracts*, the social contracts of the system are identified and formalized by means of the *contract template view diagram*	System analyst and domain expert
Verify normative context	Verify normative context	Following the guideline *Normative context verification*, the coherence among system's norms and between them and the social contracts of the system is validated	System analyst

6.2.3.3 Work Products

This section details the work products produced in this phase. A brief description of these work products is presented in Table 6.14. Following each work product is detailed and their use is exemplified by means of our running example.

Figure 6.15 describes the relation between these work products and the metamodel elements in terms of which elements are *defined* (D), *refined* (F), *quoted* (Q), *related* (R) or *relationship quoted* (RQ).

Table 6.14 Phase 3: work products

Name	Description	Work product kind
Organizational norms guideline	This guideline specifies a process to identify and formalize restrictions on the behavior of entities gained from the analysis of system requirements	Structured text
Normative document guideline	This guideline specifies a process to extract from normative documents the norms that must be implemented in a system	Structured text
Social contracts guideline	This guideline specifies a process to identify and formalize social contracts inside a specific organization	Structured text
Normative context verification guideline	This guideline analyzes how the coherence of the normative context should be verified	Structured text
Contract template view diagram	One instance of the contract template view diagram is created in order to specify each contract template	Behavioral
Organizational view diagram	The organizational view diagram, created in the previous phase of the methodology, is refined by means of adding these norms to the diagram	Behavioral
Internal view diagram	The internal view diagram of each role, created in the previous phase of the methodology, is refined by adding the social contracts and norms attached to this role	Behavioral

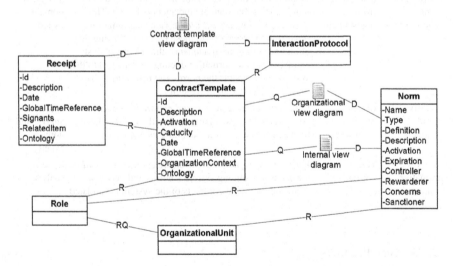

Fig. 6.15 Phase 3: relations between work products and metamodel elements

Organizational norms guideline

This guideline specifies a process to identify and formalize restrictions on the behavior of entities gained from the analysis of system requirements. These normative restrictions are associated with specific features of the system, and are usually well known by domain experts but not formally expressed in any document. This guideline helps the system analyst identify these restrictions with the support of the domain expert.

This guideline is composed of several steps that revise the requirements and specifications of the system in order to formalise the associated norms. Below, each step of the guideline is described. Figure 6.16 presents a pseudo-code algorithm that summarises these steps.

- **Step 1** *Analysis of system description documents*: These documents contain in plain text the requirements of the system, and if the system is composed by several organizations, there will be one system description document for each organization. The norms that arise from a document only affect the entities inside the organization that this document describes. The following steps are executed:

  - **1.1** Analysis of the resources: For each resource of the system, such as a database or an application, it is analyzed who has access to the resource, who cannot access it and who is responsible for its maintenance. Therefore, permission, prohibition and obligation norms are associated with these resources (lines 1–13). For example, the analysis of the *Conference* database highlight the norm that only the *chair* of the conference can modify the details of the conference (NModifyDetails, FORBIDDEN, !Chair, Modify(ConferenceDB),-,-,-,-)

  - **1.2** Analysis of the events: For each event that the organization must handle, an obligation norm to detect this event is created. If the organization should react to this event by executing a task, another obligation norm is specified (lines 18–22). The activation condition of this norm is the event itself (line 16).

  - **1.3** Analysis of the offers/demands: External stakeholders can interact with the organization, offering and demanding services or resources. If the system is obliged to offer any specific service, an obligation norm is created. If there are specific entities that are allowed to offer, demand or use a service, a related permission norm is created (lines 24–27). On the other hand, if there are specific entities that are not allowed to offer, demand or use it, a related prohibition norm is created (lines 32–35).

  - **1.4** Analysis of domain restrictions: The last attribute of the system description document analyzes if there are normative documents attached to the organization or if there are specific domain restrictions that must be taking into account. For each normative document attached to the system the guideline *Normative documents* described below must be executed (lines 38–39). If there is any restriction that is described directly by the domain expert during the analysis of the requirements of the system it must be integrated in the design (lines 40–41). For example, in the CMS case study the domain expert has claim that *"Each conference should publish a normative document describing its internal regulations"*, it

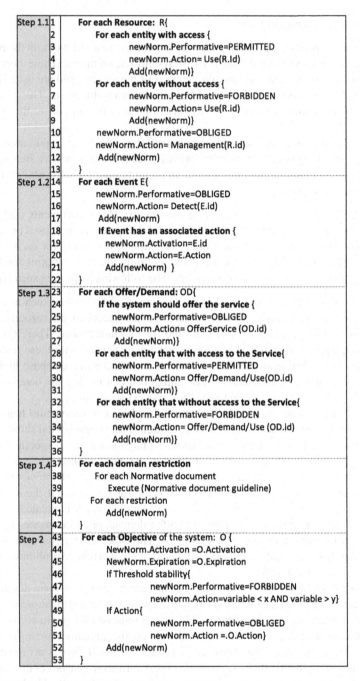

Step 1.1	1	**For each Resource:** R{
	2	**For each entity with access {**
	3	newNorm.Performative=PERMITTED
	4	newNorm.Action= Use(R.Id)
	5	Add(newNorm)}
	6	**For each entity without access {**
	7	newNorm.Performative=FORBIDDEN
	8	newNorm.Action= Use(R.id)
	9	Add(newNorm)}
	10	newNorm.Performative=OBLIGED
	11	newNorm.Action= Management(R.id)
	12	Add(newNorm)
	13	}
Step 1.2	14	**For each Event** E{
	15	newNorm.Performative=OBLIGED
	16	newNorm.Action= Detect(E.id)
	17	Add(newNorm)
	18	**If Event has an associated action {**
	19	newNorm.Activation=E.id
	20	newNorm.Action=E.Action
	21	Add(newNorm) }
	22	}
Step 1.3	23	**For each Offer/Demand:** OD{
	24	**If the system should offer the service {**
	25	newNorm.Performative=OBLIGED
	26	newNorm.Action= OfferService (OD.id)
	27	Add(newNorm)}
	28	**For each entity that with access to the Service{**
	29	newNorm.Performative=PERMITTED
	30	newNorm.Action= Offer/Demand/Use(OD.id)
	31	Add(newNorm)}
	32	**For each entity that without access to the Service{**
	33	newNorm.Performative=FORBIDDEN
	34	newNorm.Action= Offer/Demand/Use (OD.id)
	35	Add(newNorm)}
	36	}
Step 1.4	37	**For each domain restriction**
	38	For each Normative document
	39	Execute (Normative document guideline)
	40	For each restriction
	41	Add(newNorm)
	42	}
Step 2	43	**For each Objective** of the system: O {
	44	NewNorm.Activation =O.Activation
	45	NewNorm.Expiration =O.Expiration
	46	If Threshold stability{
	47	newNorm.Performative=FORBIDDEN
	48	newNorm.Action=variable < x AND variable > y}
	49	If Action{
	50	newNorm.Performative=OBLIGED
	51	newNorm.Action =.O.Action}
	52	Add(newNorm)
	53	}

Fig. 6.16 From requirements to formal norms guideline

is formalized as (confNormative, OBLIGED, Conference, Publish(ConferenceRegulations),-,-,-,-). This norm will be attached to every conference, therefore, the task of defining the internal normative should be added to a role inside the conference. In this case, this task has been added to the *chair* responsibilities.

- **Step 2** *Analysis of the objectives description document*: We can differentiate two types of objectives: the objectives attached to restrictions and the objectives attached to specific actions. First, for each objective that pursues the stability of any variable of the system in a threshold, a forbidden norm should be created to ensure that the threshold is not exceeded (lines 46–48). A variable of a system is anything that the system is interested in measuring; for example, the temperature of a room or the quantity of money that a seller earns. Second, for each objective that is attached to an action, an obligation norm is created in order to ensure that there is an entity inside the system that pursues this objective (lines 49–51). The activation and expiration conditions of the created norms are determined by the activation and expiration conditions of the related objective.

Normative document guideline

This guideline specifies a process to extract from normative documents the norms that must be implemented in a system. Normative documents can be governmental law restrictions or internal regulation of each institution. For example, any system that stores personal information must follow governmental law about personal data privacy. These documents are usually written in plain text, so they must be analysed in order to make the design of the system compliant with the norms. A system composed of several organisations or subsystems can have different normative documents associated with each party in the system. Therefore, this guideline should be employed once for each document, and the effects of the derived norms should be taken into account only inside the organisation or sub-organisation associated with the document.

This guideline presents a set of steps that help a system analyst to extract norms from normative documents. Below each step of the guideline is described. Figure 6.17 summarises the steps of guideline.

- **Step 1** *Identify matches*: This step consists of matching the actors, actions and resources addressed in the document with their related entities in the domain analysis. The binding is operated by the system analyst, comparing how entities are named in the normative document with how they are named in the domain

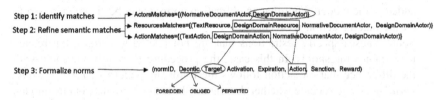

Fig. 6.17 From normative documents to formal norms guideline

analysis. The final result of this step is three sets of possible matches, one for actors, one for actions, and another for resources such as databases or applications: (1) *ActorsMatches={(NormativeDocumentActor,DesignDomainActor)}*; (2) *ActionMatches= {(TextAction, DesignDomainAction, NormativeDocumentActor, DesignDomainActor)}* (note that, if an action is associated with a specific actor either in the normative document or in the design domain, this information should be annotated); and (3) *ResourcesMatches= (TextResource, DesignDomainResource, NormativeDocumentActor, DesignDomainActor)*.

- **Step 2** *Refine semantic matches*: The domain expert should verify the semantic match of the set of identified actors, actions and resources, and should also refine the identification of actors by means of the analysis of matches between actions and resources. If an action or a resource is associated with a specific actor in the normative document and it is also associated with a specific actor in the design domain, there could be a semantic match between these actors.

- **Step 3** *Formalize norms*: When a domain actor is recognised to be a normative document subject, the corresponding rights and duties must be assigned to this actor. The normative document should be revised again and, for each identified actor, action or resource that appears in the text, a norm should be added to the design model. As explained previously, a norm is composed of the following attributes:

  - *normID*: For future maintenance activities, it is important to clearly associate each created norm with its associated clause in the normative document. Normative document clauses are usually identified by acronyms composed of a set of numbers and symbols which, in order to maintain traceability, should be used as the norm identification attribute. If the normative document clauses are not already labeled, they will be labeled the same, both in the document and in the created norms.

  - *Deontic modality and Action*: Obligations are expressed in plain text like orders, e.g. *The actor should perform the action, must, ...* The *forbidden* deontic modality is usually associated with negative sentences in which the prohibition of the execution of a task is specified. If the clause specifies the possibility of performing an action, a permission norm must be created.

    If there is a match between the action described in the normative document and an action in the design domain, the action is attached to the norm. If there is no match, then the domain expert must specify whether the action is outside the scope of the system. Only if the action is in scope and its associated deontic modality is an obligation, should it be taken into account. In this case, the specification of the requirements of the system should be revised and the action added to the model. After that an obligation norm would be created in order to ensure that the action is executed.

    Some clauses specify restrictions that are not related to actions, but related to variables thresholds. In this case, the procedure is exactly the same with the difference that the action follows the following pattern *variable* $< x$ *and variable* $> y$. To decide whether this clause should be formalised in the model, the domain expert should specify whether the variable is in scope or if any action that is performed modifies the value of this variable.

- *Target*: If the clause makes reference to an identified actor, this actor is the target of the norm. If the clause makes reference to an identified action or resource with no specific reference to any actor, the target of the norm is all members of organisations that must follow this norm.
- *Activation and Expiration*: If the clause specifies the circumstances under which it should be applied, these circumstances should be formalised as the *activation* and *expiration* attributes. Commonly, activation conditions are specified after conditional particles such as *If, When, ...* Deactivation conditions are usually specified after temporal conditional and exception particles such as *Unless, Until, Except when ...*
- *Sanction and Reward*: Sometimes the action that should be executed when a clause is fulfilled or violated is explicitly specified in the text. In such cases, it is necessary to look for a match between this action and an action from the design domain. If there is no match a new action should be added and the responsibility for executing this task should be added to one role.

Social contracts guideline

This guideline specifies a process to identify and formalize social contracts inside a specific organization regarding the information detailed in the *role description document*, the roles' *internal view digrams* and the structure of the organization. Social contracts are used to formalize two kinds of social relationship: (1) *play role contract template*, which specifies the relationship between an agent playing a role and its host organization; and (2) *social relationship contract template*, which specifies the relationship between two agents playing specific roles. Social order thus emerges from the negotiation of contracts over the rights and duties of participants.

One *play role contract template* should be defined for each role of the organization in order to establish the rights and duties that any agent playing this role should fulfill. Therefore, in the CMS case study, seven *play role contract* templates should be formalized: one for role *user* of the main organization, and six for each role described inside the *Conference* organization (*author, reviewer, PC member, Chair, Vice-chair, Publisher*). That means that the rights and duties that an agent that tries to play a role inside a conference can be different depending on how each conference negotiate these contracts. For example, one conference can establishes that a PC member cannot submit a paper to this conference while other conference do not add any restriction like that. Since every agent that intends to play a specific role inside the system must sign a *play role contract*, every agent will be aware of its rights and duties inside the organization in advance.

One *social relationship contract template* should be defined for each pair of roles that must interchange services and products as part of the social structure of the organization. Contracts of this kind should be negotiated and signed by the related entities and not by the organization as whole. However, if the terms of the contract are not negotiated by the entities, and the relationship between these agents is determined by their organization, it is not necessary to create a social relationship contract.

Instead, the rights and duties of each role over the other are included in their respective *play role contracts*. In the CMS case study there is an authority relationship between the *chair* role and the *vice-chair* role. The terms of this relationship are specified by each conference. Therefore, the rights and duties from one entity to the other are formalized in their respective *play role contract* and no social relationship contract is created.

Figure 6.18 presents a pseudoalgorithm of this guideline. Below, each step of the guideline that should be applied to each role of the system is described.

- **Step 1** *Adding identified norms*: Every restriction or norm identified during the application of the *Organizational norms guideline* that affects the role should be added to the contract (lines 3–4). The norms that are attached to several roles, but that include this specific role should be added. This can increase the size of the contract, so it is the responsibility of the domain expert to specify which norms should be communicated. For example, in the case of CMS case study, not all governmental norms related to the storage of personal data are included in the contracts, only a norm that specifies that any agent inside the system should follow this regulation is specified in the contracts.
- **Step 2** *Analysis of the organizational objectives*: In previous phases of the ROMAS methodology, the requirements of the system are analyzed by means of the analysis and decomposition of the objectives of the system. Each objective is associated with an action that must be performed in order to achieve it, and these actions are associated with specific roles that become responsible for executing them. Therefore, for each objective related to a role, an obligation norm must be created that ensures the execution of this action. The activation and expiration of the norm match the activation and expiration of the objective (lines 5–7). If the action related to the objective is a *task*, the role is obliged to execute this task (lines 8–10). If it is related to a *service*, the role is obliged to offer and register this service (lines 11–14).
- **Step 3** *Analysis of offers/demands*: The description of each role should specify which resources and services this role must offer and which ones ca use. For each resource and service that this role is able to use, a permission norm is added (lines 16–20). For each resource or service that this role cannot have access to, a prohibition norm is created (lines 21–24). Also, for each resource and service that this role is supposed to provide, an obligation norm is added (lines 26–30). In this sense, an agent would not be able to play a role unless it were able to provide all the services and resources that are specified in the *play role contract*.
- **Step 4** *Analysis of the events*: For each event that the role must handle, a norm that forces any agent that plays this role to detect this event is created (lines 31–34). If the role should react to that event by executing a task, an obligation norm is created whose activation condition is that event and indicates that the role should execute this action (lines 35–38).
- **Step 5** *Analysis of the relationships*: As is discussed above, the norms derived from the social relationships between roles should be included in the *play role contract* template when they cannot be negotiated by the entities playing these

	1	**For each Role R {**
	2	
Step 1	3	**For each Restriction** R.Rest
	4	Add(R.Rest)
Step 2	5	**For each Objective** related with the role R.O {
	6	NewNorm.Activation =R.O.Activation
	7	NewNorm.Expiration =R.O.Expiration
	8	**If R.O.Action==Task**
	9	newNorm=Performative=OBLIGED
	10	newNorm.Action=R.O.Action
	11	**If R.O.Action==Service**
	12	newNorm.Performative=OBLIGED
	13	newNorm.Action = REGISTER R.O.Action
	14	Add(newNorm)
	15	}
Step 3	16	**For each Resource or Service Used** R.RU{
	17	newNorm.Performative=PERMITTED
	18	newNorm.Action= R.RU.UseMode(R.RU.id)
	19	Add(newNorm)
	20	}
	21	**For each Resource or Service Forbidden** R.RF{
	22	newNorm.Performative= FORBIDDEN
	23	newNorm.Action= Use(R.RF.id)
	24	Add(newNorm)
	25	}
	26	**For each Resource or Service Provided** R.RP{
	27	newNorm.Performative=OBLIGED
	28	newNorm.Action= Provide(R.RP.id)
	29	Add(newNorm)
	30	}
Step 4	31	**For each Event** R.E{
	32	newNorm.Performative=OBLIGED
	33	newNorm.Action= Detect(R.E.id)
	34	Add(newNorm)
	35	**If R.E has an associated action {**
	36	newNorm.Activation=R.E.id
	37	newNorm.Action=R.E.Action
	38	Add(newNorm) }
	39	}
Step 5	40	**For each SocialRelationship** R.S {
	41	newNorm.Activation=R.S.Activation
	42	newNorm.Expiration=R.S.Expiration
	43	**If R.S.type == Incompatibility {**
	44	newNorm.Performative=FORBIDDEN
	45	newNorm.Action= PlayRole(R.S.IncompatibleRole) }
	46	**If R.S.type == ForcedCompatibility {**
	47	newNorm.Performative=OBLIGED
	48	newNorm.Action= PlayRole(R.S.compatibleRole) }
	49	**If R.S.type == (Information) {**
	50	newNorm.Performative=OBLIGED
	51	newNorm.Activation=R.S.Event
	52	newNorm.Action= Inform(R.S.SupervisorRole,R.S.Associated_Information) }
	53	**If R.S.type == (Authorization/Submission) {**
	54	newNorm.Performative=OBLIGED
	55	newNorm.Action= ProvideService(R.S.SupervisorRole,R.S.AssociatedService) }
	56	}
Step 6	57	**For each personal objective** of the role, the system can establish thresholds or any kind of limitation in their
	58	performance. These limitations will arise FORBIDDEN norms.
	59	}

Fig. 6.18 Social contracts guideline

roles, i.e. they are rigidly specified by the organization. In other cases, a *social relationship contract* should be created and these norms included in it. The norms that are derived from the social relationship should be activated only when the social relationship is active and their deontic attribute depends on the type of relationship between the parties. If two roles are incompatible, a prohibition norm is added specifying this fact (lines 43–45). In the same way, if any agent playing one role is required to play another, an obligation norm is included in the contract (lines 46–48). Usually, a social collaboration appears when several roles should interact to achieve a global goal of the organization. In such cases, a set of obligation norms specify which actions and services are the responsibility of each entity. If the collaboration relationship indicates *information*, it means that one role is obliged to inform another when some conditions occur. An *authority/submission* relationship requires the specification of: (1) which services should provide the submitted party, (2) which actions the authority can force to do to the other agent, and (3) which actions the submitted party cannot perform without the consent of the authority.

- **Step 6** *Analysis of personal objectives*: A personal objective of a role is a goal that is not directly related to the main goals of the system, but that all the agents that play this role will pursue. The system as an entity can establish some restrictions on the performance of personal objectives (lines 57–59). An example of a personal objective in the *CMS* case study is that although the agents that play the role *author* pursue the objective of *Submitting as many papers as possible*. Each conference can establish limits on the quantity of papers that an author can submit to the same conference.

Normative context verification guideline

The verification of the normative context is limited here to the verification that there are no norms in conflict, i.e., that the normative context is coherent. As is presented in [46], conflicts in norms can arise for four different reasons: (1) the obligation and prohibition to perform the same action; (2) the permission and prohibition to perform the same action; (3) obligations of contradictory actions; (4) permissions and obligations of contradictory actions. Therefore, after the specification of the organizational norms and the social contract templates that define the structure of the organization it is necessary to verify that the normative context as a whole is coherent.

Each organization can define its own normative context independently of the other organizations that constitute the system. The first step is analyzing the normative context of the most simple organizations, i.e., the organizations that are not composed by other organizations. After that, we will analyze the coherence between this simple organization and the organization in which it is inside. This process will continue until analyzing the coherence of the main organization of the system.

In order to analyze the coherence of a specific organization, it is necessary to verify that: (1) There is no state of the system in which two or more organizational norms in conflict are active. (2) There is no norm that avoids the satisfaction of an organizational objective, i.e., there is no norm that is active in the same states than the objective is pursued and whose restriction precludes the satisfaction of this objective. (3) There is no social contract that specifies clauses that are in conflict with the organizational norms. (4) There is no a pair of social contract whose clauses are in conflict between them and therefore, the execution of one contract would preclude the satisfactory execution of the other one. (5) There is no social contract in which a role participates whose clauses preclude the satisfaction of the roles objectives.

The verification task can be performed manually or by means of automatic techniques such as model checking. In [57], we present a plug-in integrated in our case tool that allows a simple verification of the coherence between the organizational norms and the contracts by means of the SPIN model checker [72].

Contract template view diagram

One contract template diagram is created for each identified social contract. The recommended steps to specify a contract template are:

- *Identify signants*: If it is a *play role contract template* the signants are the entity that tries to pursue this role and the organization as a whole. If there is a specific role in charge of controlling the access to the organization, the entity playing this role will sign the contract on behalf of the organization. If it is a *social relationship contract template* the signants are the entities playing the roles that have the relationship.
- *Attach clauses*: The norms that has been identified by means of the *social contract guideline* are included in the contract. If the norm to be included in the contract must be in any contract of this type, this norm is defined as a *hard clause*. On the contrary, if the norm to be included in the contract is a recommendation this norm is defined as a *soft clause*.
- *Define receipts*: In order to monitorize the correct execution of the contract, it is recommended to define specific artifacts that entities participating in the contract should provide in order to prove the fact that they have executed their required actions.
- *Define authorities*: Optionally, the designer can define who is responsible for verifying the coherence of the final contract (*notary*), for controlling the correct execution of the contract (*regulation authority*), and for acting in case of dispute between the signant parties (*Judge*).
- *Identify protocols*: Optionally, the designer can define specific negotiation, execution and conflict resolution protocols. At this phase, only a general description of these protocols is provided. They will be completely specify in the next phase of the methodology.

Figure 6.19 shows the *play role contract template* that any entity that wants to play the role reviewer should sign. It is signed by the role that wants to play the role *reviewer* and by the *conference* in which the entity wants to participate. There are six clauses attached to this contract template that specify an entity playing this role is not allowed to modify the details about a conference unless it is also the chair of this conference (*NModifyDetails* norm), and neither to access to the submission information about a paper in which he is also author (*Incompatibility* norm). This entity would have permission to access to the reviews database (*WriteReviews* norm) and to use the

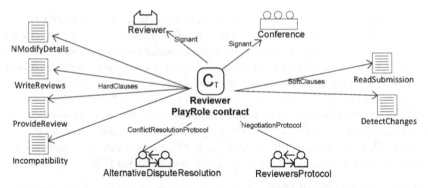

NORM ID	NORM DESCRIPTION (Deontic,Target,Activation,Expiration,Action,Sanction,Reward)
NModifyDetails	(FORBIDDEN, IChair, Modify(ConferenceDB),-,-,-,-)
WriteReviews	(PERMITTED, Reviewer, -,-, WriteAccess(Conference database),-,-)
ReadSubmission	(PERMITTED, Reviewer,-, - , UseService(Get Info Submission),-,-)
DetectChanges	(OBLIGED, Reviewer, -,-, Detect(ChangesConference event),-,-)
ProvideReview	(OBLIGED, Reviewer, -,-, ProvideService(Execute review),-,-)
Incompatibility	(FORBIDDEN, Reviewer, reviewer=author,-, UseService(Get Info Submission),-,-)

Fig. 6.19 Phase 3: case study—reviewer play role contract template

service *Get Info Submission* (*ReadSubmission* norm). This entity would be obliged
to detect when the conference details have changed (*DetectChanges* norm) and to
provide the service *Execute review* (*ProvideReview* norm).

6.2.4 PHASE 4: Activity Specification

During this phase each identified task, service and protocol is described by means
of instances of the *activity model view*.

The process flow at the level of activities is reported in Fig. 6.20. The process flow
inside each activity is detailed in the following subsections (after the description
of process roles). Figure 6.21 describes the Activity specification phase in terms
of which roles are involved, how each activity is composed of tasks, which work
products are produced and used for each task and which guidelines are used for each
task.

Fig. 6.20 Phase 4: activity tasks

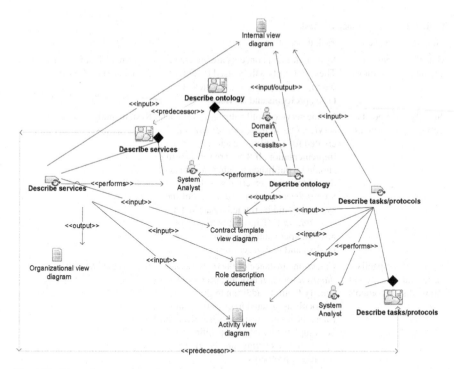

Fig. 6.21 Phase 4: resources and products used

6.2.4.1 Process Roles

The domain expert should provide the domain ontology and should give support to the system analyst in the definition of the protocols, tasks and services.

6.2.4.2 Activity Details

As Fig. 6.13 shows this phase is composed of tree activities: *Specify ontology*, *Specify services* and *Specify tasks and protocols*.

Each activity is composed of one task. All the tasks of this phase are detailed in Table 6.15. The description of the tasks details the sequence of actions that should be performed in this phase, which guidelines and work products are used by each task, and which work products are produced.

6.2.4.3 Work Products

This section details the work products produced in this phase. A brief description of these work products is presented in Table 6.16.

Table 6.15 Phase 4: activity tasks

Activity	Task	Task description	Roles involved
Specify ontology	Specify ontology	System domain concepts are analyzed. These concepts will be used to define the inputs, outputs and attributes of tasks, protocols and services	System analyst (performs) and domain expert (assists)
Specify services	Specify services	Define service profile attributes for each service. One *activity view* diagram is created for specifying each service implementation. If there are services that should be published to other members of the system or to external stakeholders, the *organizational view diagram* of the system should be refined by adding a *BulletinBoard*. This abstraction is an artifact where authorized entities can publish and search services	System analyst (performs)
Specify tasks and protocols	Specify tasks and protocols	Create one instance of the *activity view diagram* for each task and protocol to specify them. In addition to the protocols associated to objectives and roles, the contracts of the system should be completed by adding specific negotiation, execution and conflict resolution protocols	System analyst (performs)

Table 6.16 Phase 4: work products

Name	Description	Work product kind
Activity view diagram	One instance of the activity view metamodel is created for specifying every task, protocol and service implementation	Behavioral
Organizational view diagram	The organizational view diagram created in previous phases is refined in order to specify which services are published in the Bulletin Board and who have access to them	Behavioral

The flow of activities inside this phase is reported in Fig. 6.20 and the tasks are detailed in the Table 6.15. Figure 6.22 describes the relation between these work products and the metamodel elements in terms of which elements are *defined* (D), *refined* (F), *quoted* (Q), *related* (R) or *relationship quoted* (RQ).

One *activity view diagram* is created for each task, protocol or service identified in the previous phases of the methodology. Phase 2 shows the tasks, services and protocols that each role should implement and phase 3 identifies the negotiation, execution and conflict resolution protocols for the contract templates.

An example of an instance of the activity view diagram that represents a protocol is presented in Fig. 6.23. It shows the description of the *reviewer play role negoti-*

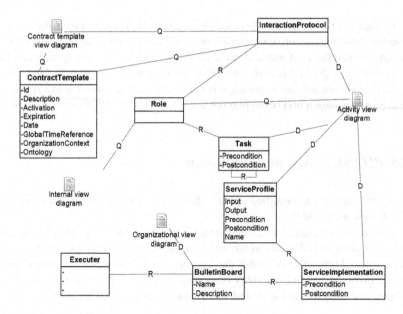

Fig. 6.22   Phase 4: relations between work products and metamodel elements

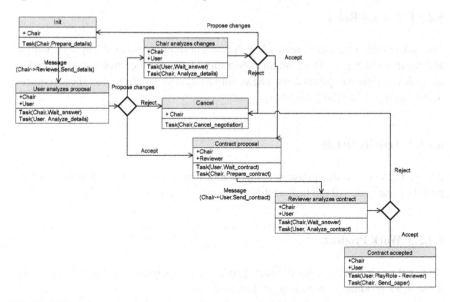

Fig. 6.23   Phase 4: case study—reviewer play role negotiation protocol

ation protocol. First, the chair sends to the user that tries to play the role reviewer the details about the conference (deadlines, topics of interests, ...). The user analyzes this information and if necessary propose a change in the review deadlines.

This change can be accepted or rejected by the chair. If the chair rejects the change, he can finish the interaction or modify his proposal and send it again to the user. Once they have agreed the conference details, the chair send the user the specification of the contract, i.e., the rights and duties that the user will acquire if he becomes a reviewer. This contract cannot be negotiated, so the user can reject it and finish the interaction or accept it and begin playing the role reviewer within this conference.

6.2.5 PHASE 5: Agents Specification

During this phase each identified agent is described by means of an instance of the *internal view* metamodel.

The process flow at the level of activities is reported in Fig. 6.24. The process flow inside each activity is detailed in the following subsections (after the description of process roles). Figure 6.25 describes the Agents specification phase in terms of which roles are involved, how each activity is composed of tasks, which work products are produced and used for each task and which guidelines are used for each task.

6.2.5.1 Process Roles

The tasks of this phase are executed by the collaboration between the system analyst and the domain expert. The domain expert should provide the information related to agent development requirements. The system analyst should formalize these requirements using the ROMAS diagrams.

6.2.5.2 Activity Details

As Fig. 6.24 shows, this phase is composed of four activities. Each activity is composed of one task. All the task of this phase are detailed in Table 6.17.

6.2.5.3 Work Products

This section details the work products produced in this phase. A brief description of these work products is presented in Table 6.18.

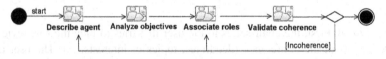

Fig. 6.24 Phase 5: activity tasks

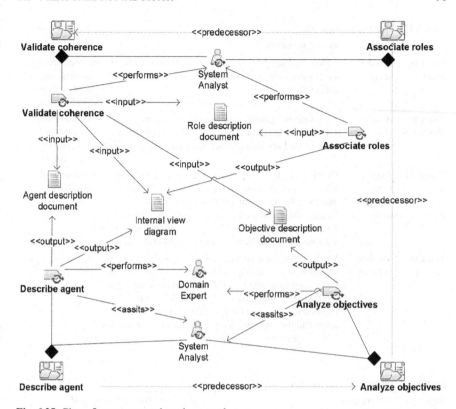

Fig. 6.25 Phase 5: resources and products used

Figure 6.26 describes the relation between these work products and the metamodel elements in terms of which elements are *defined* (D), *refined* (F), *quoted* (Q), *related* (R) or *relationship quoted* (RQ).

Figure 6.25 shows graphically the products used and produced by each task. First an *agent description document* is created for each agent. Table 6.19 shows the related guideline and an example from the CMS case study. After that, each identified objective is analyzed following the guideline *objective description document* described in Phase 1. The analysis of the objectives in our running example shows that the main objective of the PHD student agent, *Improve CV*, is decomposed in: *Submit thesis draft*, *Increase number of publications* and *Collaborate in conferences*. The first objective is not related to any objective of the system, so it cannot be achieved inside the conference management system. The second objective, *Increase number of publications*, could be achieved if the agent joined conferences as an author. The authors' play role contract template establish that any agent that wants to join a conference as an author should submit an abstract of the paper. Since Bob has unpublished papers that could submit to a conference he can play the role *author*. The third objective, *Collaborate in conferences*, could be achieve by being the PC

Table 6.17 Phase 5: activity tasks

Activity	Task	Task description	Roles involved
Describe agent	Describe agent	Following the guideline *agent description document*, the development requirements of each agent are analyzed	System analyst (assists) and domain expert (performs)
Analyze objectives	Analyze objectives	Following the guideline *objectives description document* detailed in Phase 1, the agent's objectives are analyzed and decomposed in operational objectives	System analyst (assists) and domain expert (performs)
Associate with system roles	Associate with system roles	Identify which roles the agent must play in order to achieve its objectives. This analysis is performed by matching the agent objectives with the roles functionality. Therefore, the *objective description document* of the agent is compared with the analysis of the roles presented in the *roles description documents*	System analyst (performs)
Validate coherence	Validate coherence	Validate that the normative context of the agent does not avoid any of its objectives to be satisfied. Validate that the agent is able to fulfill its commitments defined by its signed contracts. Validate that there is no incoherence between the normative context of the agent and the normative context of the organizations to which it pertains	System analyst (performs)

Table 6.18 Phase 1: work products

Name	Description	Work product kind
Agent description	This document analyzes the main features of each agent and its relationship with the system	Structured text
Objectives description	This document analyzes the individual objectives of each agent and decomposes them into operational objectives. This document is the same document that the one used in the first phase of the methodology to analyze the system's objectives	A composite document composed by a structured text document and a diagram
Internal view diagram	One internal view diagram is created for each agent. They graphically represent the specification of each agent	Behavioral

member of a conference. However, after the validation step it is shown that Bob cannot play the role PC member because any agent that wants to play this role must be a doctor and the agent is a PHD student. One *internal view diagram* is created for each to specify the features of each agent. As an example, Fig. 6.27 shows the internal view diagram of the PHD student agent.

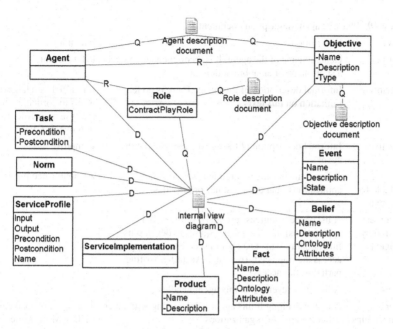

Fig. 6.26   Phase 5: relations between work products and metamodel elements

6.3 Work Product Dependencies

According to the FIPA standard *Design Process Documentation Template* this section provides a representation of the dependencies among the workproducts generated during the development process (Fig. 6.28). The direction of the arrow points from the input document to the consumer one. For instance, the analysis of the *system description document* is used as an input to specify the *objective description document* and the *use case diagrams*.

6.4 Conclusions

This chapter presents the ROMAS methodology. ROMAS is a new methodology that deals with the analysis and design of normative open MAS. This methodology proposes a metamodel that integrates the concepts of agents, organizations, services, norms and contracts. By means of these high-level abstraction concepts ROMAS generates designs that are very close to real-life systems and therefore, that are easily understood by domain experts. ROMAS metamodel allows specifying the global system requirements and objectives, as well as, the individual features and requirements of every entity.

Table 6.19 Phase 5: agent description document

Property	Description	Example
Identifier	General name of the agent. It is recommended to select a short name or an abbreviation	PHD student
Description	Informal description of the agent. There is no length limitation on this text	It is a PHD student who wants to participate in the system in order to improve its CV
Objectives	Informal description of the agent's purposes of the agent	• Improve its CV
Resources: available for the agent	A list of the resources (products, services and applications) that the agent has or provides	• Unpublished papers
Resources: required by the agent	A list of the resources (products, services and applications) that the agent requires to develop its functionality. This text should specify which type of access the role needs (reading, executing, writing, partial or full access)	
Events	A list of the events that this agent handles	
Other memberships	A text explaining if this agent is interacting with other active systems or organizations	This agent plays the role PHD student inside its college
Restrictions	A list of the restrictions that are inherent to the agent	This agent must follow the regulation of its college and that he is responsible of the maintenance of the research group database

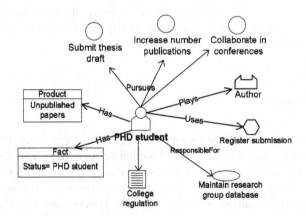

Fig. 6.27 Phase 5: case study—PHD student agent description

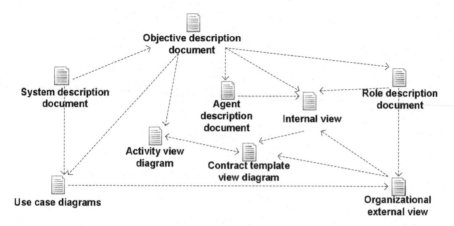

Fig. 6.28 Work product dependencies

The normative context of the system is specified by means of norms and contracts. This fact gives flexibility in the way each individual entity achieve its objectives inside the system at the same time that ensures the stability of the system. Thanks to the structure of the system by means of organizations, different normative contexts can be specified. This is an interesting feature since we are dealing with autonomous entities and institutions that can have their own specific regulations attached.

ROMAS methodology specifies a development process composed by a set of activities and tasks that guide developers at each step of the process. ROMAS offers a set of guidelines that facilitate the decision-making process at critical points such the identification and formalization of the normative contexts of the system, the decomposition of the functionality of the system in roles, the verification of the coherence of the normative context and so on.

Chapter 8 evaluates this methodology regarding the evaluation criteria presented in Chap. 2 and analyzes its contributions to the current state of the art.

Chapter 7
ROMAS Development Framework

This chapter presents the ROMAS development framework that offers support for the application of the ROMAS methodology. This framework has been implemented using model-driven technology to integrate the analysis, design, verification and code generation of normative open MAS.

The rest of the chapter is organized as follows: Sect. 7.2 introduces the technological background of this framework. Section 7.3 explains how to use this framework to develop normative open MAS. Section 7.4 details the functionalities of the modeling tool. Section 7.5 introduces the proposed prototype for verifying the modeled designs and discusses how this prototype could be improved.

7.1 Motivation and Objectives

The ROMAS methodology offers a set of guidelines and specifies a development process for analyzing and designing normative open MAS. However, in order to become of practical use, tools that support this methodology should be provided. The main objective of the ROMAS development framework is to integrate the analysis, design, verification, and code generation of normative open systems.

Our first objective is the creation of a CASE tool that facilitates the graphical representation of the diagrams generated by the ROMAS methodology. This fact guides the challenge of enabling the modeling of different diagrams whose entities and relationships are restricted by the different views of the ROMAS metamodel.

Our second objective is the verification of the coherence of the designed normative context. Normative open systems may have to integrate different normative contexts derived from each organization involved in the system. Besides, the social contracts and contractual agreements specified in the system should be coherent between them and with the normative context of the system. An incoherence in the specification of the normative context may produce critical situations as well as a lack of robustness and trustworthiness. Therefore, it is important to detect and correct these potential incoherences at an early stage of the development process.

© Springer International Publishing Switzerland 2015
E. Garcia et al., *Regulated Open Multi-Agent Systems (ROMAS)*,
DOI 10.1007/978-3-319-11572-6_7

Our third objective is to allow the automatic generation of code from the ROMAS designs. Automatic code generation avoids implementation mistakes and reduces implementation time. As we are dealing with normative open systems that can include several institutions, entities, norms, and contracts, the implementation task can be complex and it is important to offer techniques that reduce this complexity.

The ROMAS methodology has been specified using a FIPA standard to facilitate the comparison of methodologies to reduce learning time and also to facilitate the extraction and introduction of methods fragments. In that sense, the ROMAS methodology is open to integrate guidelines and methods fragments from other methodologies. Therefore, the ROMAS development framework should be easily extensible and interoperable with other methods and tools.

To achieve these objectives and integrate the multi-agent, we use model-driven techniques to create a flexible and extensible development framework that includes the analysis, design, verification, and code generation of normative open systems.

7.2 Technology Background: Model-Driven Architecture and Eclipse Technology

Recently, Model-Driven Development (MDD) has been recognized and has become one of the major research topics in the agent-oriented software engineering community due to its inherent benefits [81]. Its use improves the quality of the developed systems in terms of productivity, portability, interoperability, and maintenance [4, 65]. Basically, MDD proposes an approach to software development based on modeling and on the automated mapping of source models to target models. The models that represent a system and its environment can be viewed as a source model, and code can be viewed as a target model. In this sense, MDD aims to change the focus of software development from code to models. This paradigm shift allows developers to work with the high abstraction level inherent to multi-agent systems during the analysis and design stage and to transform these designs into final code in an easy way.

Some works like [4, 27, 52, 96, 100, 124] show how MDD can be effectively applied to agent technologies. Furthermore, they show how the MDD technology ca help to reduce the gap between the analysis phase and the final implementation.

The Model-Driven Architecture initiative (MDA) [106] has proposed a standard for the metamodels of the specification languages used in the modeling process known as the Meta Object Facility (MOF). This includes a set of requirements for transformation techniques that are applied when transforming a source model into a target model. This is referred to as the Query/View/Transformation (QVT) approach [3].

Following these MDA standards, the Eclipse Platform [1] is an open-source initiative that offers a reusable and extensible framework for creating IDE-oriented tools. The Eclipse Platform itself is organized as a set of subsystems (implemented in one

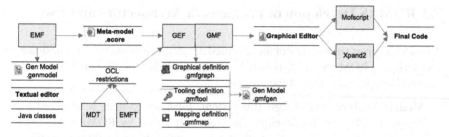

Fig. 7.1 Eclipse plugin structure

or more plug-ins) built on top of a small runtime engine. Plug-ins define the extension points for adding behaviors to the platform, which is a public declaration of the plug-in extensibility. Figure 7.1 shows the plug-ins used to develop the ROMAS development framework:

- The Eclipse Modeling Framework (EMF) plug-in offers a modeling framework and code generation facility for building tools and other applications based on a structured data model. From a metamodel specification described in XMI, Rational Rose, or the ECore standard (a variant of the MOF standard), EMF provides tools and runtime support to produce a set of Java classes for the model. EMF also provides the foundation for interoperability with other EMF-based tools and applications. Moreover, EMF generates a **textual modeler editor** from the metamodel

- The Graphical Editing Framework (GEF) and Graphical Modeling Framework (GMF) plug-ins allow developers to create a rich **graphical editor** from an existing ECore metamodel. These plug-ins allow the definition of graphical elements that are to be used in the generated tool. They also allow the definition of several views of the model and the palette of elements of each view. Finally, these plug-ins combine the graphical definition with the metamodel elements and with the different views of this metamodel, creating a complete modeling tool. These new tools are integrated into the platform through plug-ins that allow the definition of models based on the specification of the metamodels.

- The Xpand plug-in offer a language to define matching rules between the ECore metamodel and another language. A plug-in generated using Xpand consists of a set of transformations mapping rules between the entities and relationships of a metamodel defined in the ECore language and any other description language. These scripts are executed on an instance of the metamodel, i.e., on a user application model. These scripts have access to each entity and relationship of the model and match this information with the mapping rules defined at metamodel layer to generate the related code. Therefore, users can design their models using the graphical editor and execute these rules to **automatically generate code** from these models.

7.3 ROMAS Development Framework Architecture and Use

This section gives an overview of the proposed development framework for designing and verifying ROMAS which is based on a Model-Driven architecture. Figure 7.2 summarizes the main steps of the process.

1. **Model: Analyze and design the system**. First, the system is analyzed and designed following the development process specified by ROMAS methodology. During the phases of the methodology, a set of diagrams that are instances of the ROMAS metamodel are generated. These diagrams are represented graphically by means of the Eclipse modeling tool described in Sect. 7.4. This tool has been developed following the MDA [106] standards by means of the Eclipse technology. It consists of several Eclipse plug-ins that offer several graphical editors (one for each view of the ROMAS metamodel). The entire information detailed in the different diagrams is saved in a single ecore model. Therefore, all the diagrams of the same model are connected, and designers can navigate from one view to another by clicking on the main entity of the diagram. In this way, a system modeled with the ROMAS tool consists of a single ecore model and a set of graphical diagrams developed with different graphical environments.

2. **Verification of the model**. This step of the process consists of verifying the correctness, completeness, and coherence of the designed model. Although the modeling tool restricts the model to the syntax defined in the metamodel, many conflicts such as the coherence between agents' goals and the goals of their organization can arise. The current version of the ROMAS tool deals with conflicts

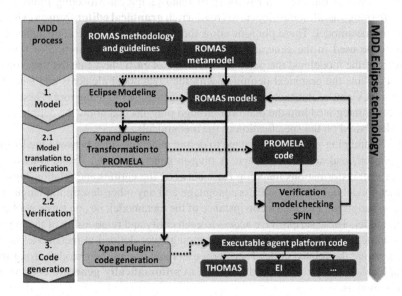

Fig. 7.2 ROMAS development framework architecture

related to the incoherences between the designed contract templates and the organizational norms. Model checking techniques are used to verify the models. The verification plug-in and process are detailed in Sect. 7.5.

3. **Generate the code for the execution platform.** Finally, the models generated by the ROMAS modeling tool can be translated into executable code. The Xpand plug-in is used to generate several patterns that transform the model designed with the modeling tool into a source code of an agent platform. The ROMAS framework architecture would be able to generate code for any agent platform as long as the transformation patterns are encoded using the Xpand plug-in. Currently, there is ongoing work [98] to create an automatic code generation plug-in for the Thomas platform [29]. Thomas is an agent platform that supports the description of multi-agent systems in social and normative environments. The details of the implementation of this plug-in are beyond the scope of this book.

7.4 ROMAS Modeling Tool

The ROMAS tool is a CASE tool for developing normative open MAS using ROMAS methodology. The development of this tool is an ongoing work. A prototype can be downloaded from http://www.gti-ia.upv.es/sma/tools/ROMAS/index.php.

The ROMAS tool derives from our previous work, the EMFGormas tool [53]. EMFGormas supports the development of organizational systems by means of GORMAS methodology. The EMFGormas tool has been modified to support the ROMAS metamodel.

7.4.1 ROMAS Tool Technical Details

The ROMAS tool is a combination of tools based on the Eclipse Modeling Framework (EMF) and tools based on the Graphical Modeling Framework (GMF) integrated into a single editor. Developed as an Eclipse plug-in, ROMAS tool is fully open source and follows the MDD principles of tool development, as Fig. 7.2 shows.

The implementation of the ROMAS tool has been performed following the MDA standards [106] by means of the Eclipse technology. The sequence of Eclipse plug-ins used to implement this tool is presented in Fig. 7.1.

First, the ROMAS metamodel was codified using the ecore standard. The specification of a metamodel using such standard makes the final plug-in interoperable with other plug-ins and tools. The ecore specification of the metamodel is extended with the default EMF edit and editor plug-ins to provide model accessors and the basic tree-based editor for the creation and management of ROMAS models. Figure 7.3 shows a snapshot of this textual editor. One ecore file is created for each system modeled with ROMAS. These files store the entities and relationships that define the

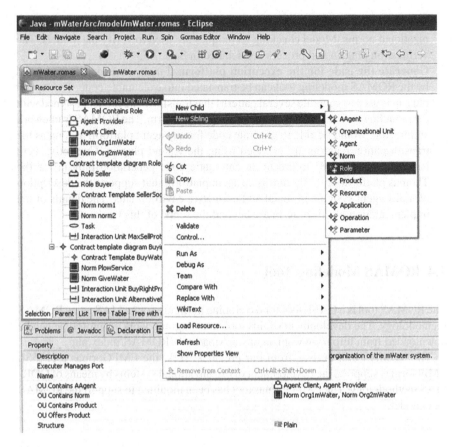

Fig. 7.3 ROMAS textual editor: mWater case study

system model. Models can be edited directly modifying its ecore file (Fig. 7.4) or using the textual editor facilitated by Eclipse (Fig. 7.3).

However, from the final user point of view this textual editor is not enough. Modeling a system using the textual editor can be challenging and time-consuming. For this reason, this basic textual editor was extended with graphical interfaces for editing the ROMAS models. As Fig. 7.1 shows these graphical interfaces were created by means of the GEF and GMF plug-ins. The ROMAS tool provides four graphical editors, one for each view of the ROMAS metamodel. Each editor restricts the modeling task to the elements and relationships specified in the corresponding view of the metamodel.

ROMAS combines these editors into a single package. The entire information detailed in the different graphical diagrams is saved in a single ecore model. Therefore, all the diagrams of the same model are connected, and designers can navigate from one view to another by clicking on the main entity of the diagram. In this way,

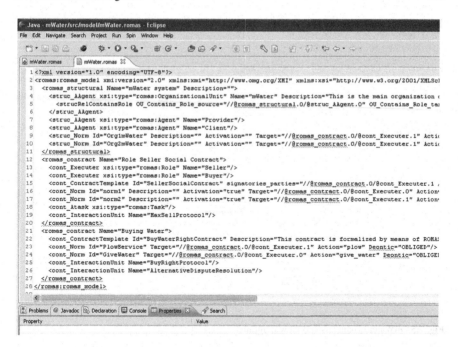

Fig. 7.4 Xml view of the mWater ecore case study

a system modeled with the ROMAS tool consists of a single ecore model and several diagrams developed with different graphical environments.

Figure 7.5 shows the user interface of the ROMAS tool. The interface has five main components: the *Eclipse Project navigator* that shows the hierarchy of project files; the *Diagram editor* where the diagrams are drawn; the *Editor palette* that contains the entities and relationships that can be modeled in the selected view of the metamodel to be selected; and the *Properties view* where the attributes of each entity and relationship are managed.

Each graphical editor provides its own editor palette which show the entities and relationships that are allowed in its corresponding view of the ROMAS metamodel.

The graphical notation used to represent each entity of the metamodel is the same as that presented in Sect. 5.2. It is summarized in Fig. 5.5.

7.4.2 Use of the ROMAS Modeling Tool

Since the ROMAS modeling tool is implemented as a set of Eclipse plug-ins, its installation is simple. It needs as basis the Eclipse modeling tool for your specific operating system. The installation process of the ROMAS tool is to copy the ROMAS plug-ins into the Eclipse plug-ins folder.

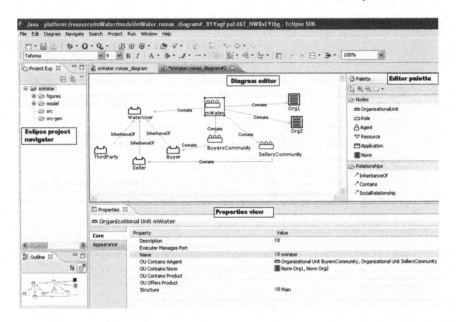

Fig. 7.5 mWater organizational view diagram

To model a system using the ROMAS tool, first the user needs to create an Eclipse project. Next, the user adds a new ROMAS diagram. This action will create an ecore file where the model will be stored as text and a diagram file that allows to graphically model the system. From this diagram file, the user can create as many instances of the metamodel views as required. Each instance is edited with its corresponding graphical editor.

Users can navigate between diagrams by means of double-clicking in the main entity of the diagram. Entities that has been created in one diagram can be reused in the others by use of shortcuts.

ROMAS modeling tool offers a traditional interface and menus. This fact improves the usability of the system and reduces the learning time.

7.4.3 Contributions and Limitations

The ROMAS tool effectively supports the analysis and design of normative open MAS following the ROMAS methodology. It offers textual and graphical editors that restrict the models to the entities and relationships specified in the ROMAS metamodel.

The ROMAS tool offers an interface that is similar to any other CASE tool and that follows the Eclipse standards. This fact increases its usability and reduces the learning time.

The ROMAS tool stores the models following the ecore standard. This fact facilitates the reusability of the models. For example, an agent could read the structure of the system of the model and reason about it at runtime.

The implementation of the ROMAS tool as Eclipse plug-ins allows to integrate this tool with other Eclipse plug-ins that can extend its functionality.

However, during the design of our case studies using the ROMAS tool we also detected some open issues and drawbacks.

Currently, the ROMAS tool supports the specification of instances of the ROMAS metamodel. However, we are still working to offer support to the specification of the textual work products that the use of the methodology produces such as the *system description document*.

The current version of the ROMAS tool does not integrate the guidelines offered by the methodology. If the modeling tool were to integrate all the work products and guidelines of the methodology, the modeling tool would be able to guide developers step-by-step the methodology. This fact will reduce the learning time of the methodology and increase the usability of the modeling tool.

Another issue that is common in Eclipse modeling tools is that it is not possible to easily reuse parts of one model into another. In the development of MAS, it is common to reuse interaction protocols or social structures between projects. So, it would be very useful to be able to import and export parts of the model.

7.5 ROMAS Module for Formal Verification

Validating that the designed system fulfills all the requirements identified in the analysis stage and verifying the coherence and completeness of such designs are common open issues in any software development approach. It has even greater importance in the development of normative open systems, where it has two specific features. First, systems of this kind integrate the global goals of the system with the individual goals of each party, where these parties are completely autonomous and their interests may conflict. It is thus crucial in helping developers to verify that the combined goals of the parties are coherent and do not conflict with the global goals of the system. If any incoherence is detected, the developer should be able to determine when this will affect the global goals and whether it is necessary to introduce norms to avoid related problems. Second, such systems usually integrate different normative contexts from the different organizations involved, which must be coherent with the contracts defined in the system. In this respect, an open question is how consistency and coherence of norms and contracts can be automatically checked in an organization.

In this section, we present a set of plug-ins that is integrated into the ROMAS tool to verify the coherence of the normative context of the systems designed using this CASE tool. As presented in Sect. 7.2, model checking techniques allow the formal verification of systems. Section 7.5.2 details our approach for verifying the normative context by means of the model checking techniques.

7.5.1 Related Work

Model checking is an area of formal verification that is concerned with the systematic exploration of the state spaces generated by a system. Model checking was originally developed for the verification of hardware systems, and has been extended to the verification of reactive systems, distributed systems, and multi-agent systems.

Although there are some works that apply model checking techniques to the verification of contract-based systems, it is still an open research topic. Some works, like that by Solaiman et al. [105], model contracts as a finite automata that models the behavior of the contract signatories. Other works, like Hsieh et al. [76], represent contracts as Petri nets. These representations are useful to verify safety and liveness properties.

The use of deontic clauses to specify the terms of a contract allows conditional obligations, permissions, and prohibitions to be written explicitly. Therefore, they are more suitable for complex normative systems like ROMAS. Work by Pace, Fenech et al. [94, 46] specifies a deontic view of contracts using the *CL* language. Pace et al. [94] use model checking techniques to verify the correctness of the contract and ensure that certain properties hold. While Fenech et al. [46] present a finite trace semantics for *CL* that is augmented with deontic information as well as a process for automatic contract analysis for conflict discovery. In the context of service-oriented architectures, model checkers have recently been used to verify compliance of Web service composition. One example is the work by Lomuscio et al. [85] that presents a technique based on model checking for the verification of contract-service compositions.

In the context of verification techniques for MAS, there are some important achievements using model checking. Walton et al. [118] use the SPIN model checker to verify agent dialogs and prove properties of specific agent protocols such as termination, liveness, and correctness. Bordini et al. [14] introduce a framework for the verification of agent programs. This framework automatically translates MAS that are programmed in the logic-based agent-oriented programming language AgentSpeak into either PROMELA or Java. It then uses the SPIN and JPF model checkers to verify the resulting systems. Work by Woldridge et al. [120] proposes a similar approach, but it is applied to an imperative programming language called MABLE. Nardine et al. [93] verify the compatibility of interaction protocols and agents deontic constraints. However, none of these approaches are suitable for ROMAS as they do not consider organizational concepts.

There are only few works that deal with the verification of systems that integrate organizational concepts, contracts, and normative environments. The most developed approach is presented in the context of the IST-CONTRACT project [92]. It offers contract formalization and a complete architecture. It uses the MCMAS model checker to verify contracts. However, to our knowledge, it does not define the organizational normative context or verify the coherence of this context with the contracts.

The ROMAS development framework tries to provide a different approach for verifying ROMAS. It is distinct in that it designs and offers a module that allows:

(1) the explicit formalization of social and commercial contract templates at design time; (2) the automatic translation of contract and norm descriptions into a verifiable model checking language; (3) the verification at design time of whether a contract template contradicts the designed normative and legal environment.

7.5.2 Verifying the Coherence of the Normative Context

The ROMAS tool integrates a set of Eclipse plug-ins to verify that there is no conflict between the organizational norms, agent norms, and contract templates designed. This verification task is associated to the last task of the *Normative context specification* phase of the ROMAS methodology.

As presented in [46], conflicts in contracts and norms can arise due to four different reasons: (1) the obligation and prohibition to perform the same action; (2) the permission and prohibition to perform the same action; (3) obligations of contradictory actions; (4) permissions and obligations of contradictory actions. At the moment, ROMAS tool verifies the first and second conflict scenarios. The last two scenarios need semantic analysis of the ontology which is part of our future work.

In order to perform the model checking verification, we use SPIN model checker [72]. The reasons to choose this model checker are that SPIN is a popular open-source software tool used by thousands of people worldwide for formal verification of distributed software systems. The software has been available freely since 1991, and continues to evolve to keep pace with new developments in the field. In April 2002, the tool was awarded the prestigious System Software Award for 2001 by the ACM. Moreover, there is an open-source Eclipse plug-in that integrates the SPIN model checker into any Eclipse application [83].

Therefore, a ROMAS verification process is executed in two steps:

1. *Prepare the model to be verified.* The modeled system is translated into a language that can be verified using model checking.
   As explained in Sect. 7.4, the systems designed using ROMAS modeling tool are stored in an ecore file. Since we use the SPIN model checker, the ecore model is translated into the PROMELA language and Linear Temporal Logic (LTL) formulas. This translation is automatically performed by the Eclipse plug-in ROMAS to PROMELA code transformation (RO2P) that is detailed in the following subsection (*RO2P*).
2. *Execute the model checker.* Once the PROMELA file and the LTL formulas have been generated, the SPIN formal verification of the model is directly run from the modeling tool. This is possible, thanks to the Eclipse plug-in[1] [83] that has been integrated into our Eclipse modeling tool. After verification, if there is any incoherence the designer must revise the model, and the verification step begins again. Figure 7.6 shows the interface of the SPIN Eclipse plug-in.

---

[1] http://lms.uni-mb.si/ep4s/.

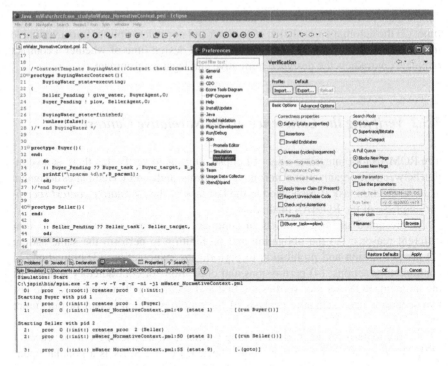

Fig. 7.6 Eclipse plug-in for SPIN interface

7.5.3 ROMAS to PROMELA Code Transformation (RO2P)

The coherence of the normative context can be verified at different levels: (1) Organizational level, where we verify that the normative context of the organizations involved in the system are coherent between them; (2) Role level, where we verify that the norms and contracts related to the roles of an organization are coherent; (3) Agent level, where we verify that the norms and contracts related to an agent that is playing a set of roles are coherent between them.

Therefore, the elements of the model that need to be translated into a verifiable code are:

- *Organizational norms*: Norms that are associated to a specific organization.
- *Role norms*: Norms that are associated to a specific role.
- *Agent norms*: Norms that are associated to a specific agent.
- *Social contracts*: Contract templates that can specify social relationships between roles or the restrictions that an agent needs to fulfill in order to play a specific role.
- *Contractual agreements*: Contract templates that specify resources or products interchanges.

Multi-agent systems have some properties that need to be considered when performing the verification:

- *Concurrency*: The entities of the system are running in the same environment concurrently. Besides, several contracts may need to be executed at the same time.
- *Dynamicity*: We are dealing with dynamic systems where norms and contracts can be activated or disactivated depending on environmental conditions, the result of other tasks, or the internal state of the system.
- *Autonomous entities*: The entities of the system are autonomous, so they have their own rational process to decide the execution task order. So, if for example an agent has signed two contracts that force him to perform two actions, we cannot be sure which action is going to be executed first. Therefore, it is necessary to analyze all the possible combinations to ensure that norm is violated.

The following section details how the ROMAS model is translated into the SPIN model checking language. Here, only a brief overview of the translation process is presented.

- Each organization, role, or agent (depending at which level is the verification performed) is translated as an individual process. Each process has a channel associated where all the tasks that this entity has to perform are stored. Each process is independent of the other and simulates the execution of the tasks that are in its channel.
  Each process is independent of the others and they can be executed concurrently. They choose randomly the next task to execute in order to simulate the autonomy of each entity in terms of which action executes first.
- Each contract with obligation or permitted norms associated translates as another process. This process is activated only in the meantime the activation condition of the contract has occurred and the expiration condition of the contract has not occurred. A contract process adds to the channels of the corresponding entities the tasks that are associated to the obligation and permitted norms.
- Each prohibition norm associated to a contract or directly to an evaluated entity is translated into an LTL property. If several prohibition norms have to be verified at the same time they are joined into only one LTL, since the SPIN model checker only permits the verification of one LTL at a time.

Translating each entity as an independent process, we ensure the concurrency of the system. Translating each contract as an independent process allows to activate and deactivate them regarding its activation and deactivation conditions. The use of channels allows the simulation of the execution of tasks. The next task to be executed is chosen randomly to simulate the capacity of each entity to decide at each moment which task to execute next.

Fig. 7.7 mWater organizational view simplified diagram

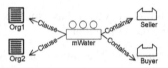

(*) (Org1, true, false,ANY,FORBIDDEN,pay_water(t,p,q)&& (p>0'5||p<0'001), -,-)
(*)(Org 2,true,false,Buyer,FORBIDDEN,plow(t,q),-,-)
(*)Norms:{Id,Activation,Expiration,Target,Deontic,Action,Sanction,Reward}

7.5.3.1 Case Study

To obtain a clear scenario to show how a ROMAS model is verified with the ROMAS CASE tool, a simplified scenario of the mWater case study is presented [64]. This system is a water market that is an institutional, decentralized framework where users with water rights are allowed to voluntarily trade their water rights fulfilling some preestablished rules. An introduction of this case study is presented in Sect. 9.2 and a complete description of this case study modeled with ROMAS is presented in [59].

- *mWater Organizational view*: Figure 7.7 shows a simplification of the *mWater organizational view* diagram where only the entities and relationships that are related to the verification of the normative context are shown. There is an organization called *mWater* that contains two types of roles *Sellers* and *Buyers*. This organization restricts the behavior of these roles by means of two organizational norms:

  - *Org1 mWater norm*: It specifies that there is a minimum and maximum price for the water, i.e., it is forbidden to pay more than 50 euro/kL or less than 1 euro/kL.
  - *Org2 mWater norm*: It specifies that an agent who is playing the *Buyer* role cannot offer services to other agents in exchange of water. This rule forces that agents only can transfer water rights with money.

- *Buying water rights contract*: Figure 7.8b represents the contract template that indicates the restrictions that any contract for buying water rights should fulfill. A contract of this type should have two signatory parties (one entity playing the role *Buyer* and other playing the role *Seller* inside the *mWater* organization. The clauses specify that the Buyer should plow the field of the Seller in exchange for a water right. The formal specification of each norm is presented in Fig. 7.8a.
- *Role Seller Social Contract template*: Figure 7.8b represents the contract template that indicates the restrictions that any contract between the organization mWater and an agent that wants to play the role Seller should fulfill. In other words, any agent who wants to play the *Seller* role must sign a contract compliant with this contract template. The norm associate to this contract template indicates that the final contract should specify the maximum number of liters of water that this agent can sell (Norm *MaxLiters*). The number of liters is defined at runtime during the negotiation between the agent and the organization.

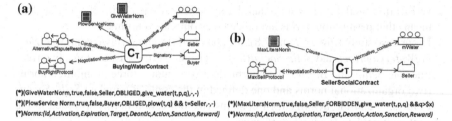

(*)(GiveWaterNorm,true,false,Seller,OBLIGED,give_water(t,p,q),-,-)
(*)(PlowService Norm,true,false,Buyer,OBLIGED,plow(t,q) && t=Seller,-,-)
(*)Norms:(Id,Activation,Expiration,Target,Deontic,Action,Sanction,Reward)

(*)(MaxLitersNorm,true,false,Seller,FORBIDDEN,give_water(t,p,q) &&q>$x)
(*)Norms:(Id,Activation,Expiration,Target,Deontic,Action,Sanction,Reward)

Fig. 7.8 **a** Buying Water Contract template **b** Role Seller Social Contract template

```
 1  «IMPORT romas »
 2  «DEFINE root FOR romas_model»
 3      «REM» /********  VARIABLES  **************/ «ENDREM»
 4      «LET (List[ContractTemplate]) {} AS contractList »
 5      «LET (List[Executer]) {} AS executerList »
 6      «LET (List[Norm]) {} AS nForbiddenList »
 7      «LET (List[Norm]) {} AS nObligedList »
 8      «REM» /********  Prepare the lists **********/ «ENDREM»
 9      «EXPAND fillLists(contractList, executerList,nForbiddenList,nObligedList) FOR this»
10      «REM» /********  Translate Forbidden Norms *********/ «ENDREM»
11      «FILE "LTL_Norms" + ".pml" »
12          «EXPAND writeForbiddenNorms FOR nForbiddenList »
13      «ENDFILE »
14      «FILE "PromelaFile" + ".pml" »
15          «EXPAND writeContracts FOR contractList »
16          «EXPAND writeExecuters FOR executerList »
17          «EXPAND writeInit(contractList, executerList)»
18      «ENDFILE »
19  «ENDLET » «ENDLET » «ENDLET » «ENDLET » «ENDDEFINE»
```

Fig. 7.9 Xpand script: main routine

7.5.3.2 Technical Details

ROMAS to PROMELA code transformation (RO2P) has been developed as an Eclipse plug-in based on the Xpand language of the Model to Text (M2T) project [2]. As illustrated in Fig. 7.1, Xpand helps developers to translate models that are defined using the ecore standard into other languages. Xpand scripts apply transformation patterns to the entities and relationship of the model and generate the related code.

This section details the Xpand transformation mapping rules between the ROMAS metamodel and the SPIN verifiable language (PROMELA code and LTL formulas). Since this plug-in is based on Xpand, it consists of a routine that specifies a set of mapping rules defined at metamodel layer. The main routine of the RO2P plug-in is presented in Fig. 7.9 and is described below.

- Figure 7.9, lines 4–7 create a set of lists to save the contract templates and norms of the system, and the entities that are affected by these norms or contracts (agents, roles, and organizations).
- Figure 7.9, line 8 invokes the *fillLists* routine that navigates the model and initialize the lists defined above. The deontic attribute of the norms indicates that they can be obligations, permissions, or prohibitions. Norms of permission can only produce a conflict if there is a prohibition over the same action. Therefore, to create the

verification model, we assume that the agent actually performs the action. This means that permission norms are modeled as obligation norms.

In our case study (Figs. 7.7 and 7.8), the *contractList* contains two contracts (*BuyingWaterContract* and *SellerSocialContract*), the executerList contains two roles (the role *Seller* and *Buyer*), the nForbiddenList contains three prohibition norms (two organizational norms and one defined in the contract template *SellerSocialContract*), and the nObligedList conatins two obligation norms specified in the contract template *BuyingWaterContract*.

- Figure 7.9, lines 11–13 invoke the routine *writeForbiddenNorms* in order to translate the norms whose deontic attribute indicates prohibition as LTL formulas. These formulas are saved in a file called *LTL_Norms.pml*. The code of the routine *writeForbiddenNorms* is presented in Fig. 7.10. Figure 7.10 line 68 creates a LTL formula that indicates that there never occurs a forbidden action. This LTL also adds the activation and expiration condition of the norm in order to perform this verification only when the norm is active. For example, the norm Org2 from the mWater organizational diagram which is formalized as: (Org2,true,false,Buyer,FORBIDDEN,plow(t,q),-,-) considering the following syntax: (Id, Activation, Expiration, Target, Deontic, Action, Sanction, Reward), is translated to LTL as: Org2 []!(Buyer_task==plow). This norm does not have activation or expiration conditions, so the LTL formula only expresses that it is not possible that a *Buyer* executes the task *plow*.

- Figure 7.9, lines 14–18 create the file *PromelaFile.pml* where the PROMELA code is saved by means of the execution of the following routines:

  - *writeExecuters routine* (Fig. 7.11): Each entity (agents, roles, organizations) is represented by an active process (line 119). The core of this process is a loop that checks its pending tasks and simulates its execution. Each party stores its pending tasks in a channel, which is a global variable accessible for all the processes (line 113). If an agent is obliged to execute a service, it is supposed to do that. Thus, the action of the obligation norm is added to the channel of the corresponding agent. This agent will simulate the execution of all the action of its channel. As an example, Fig. 7.12 shows the PROMELA code for the role *Buyer* of the case study.

  - *writeContracts routine* (Fig. 7.13): Each contract template is specified as a PROMELA process (Fig. 7.13, line 83). As an example, Fig. 7.14 shows the PROMELA code for the *BuyWaterRightContrat template* whose ROMAS diagram is shown in Fig. 7.8a.

```
65   «DEFINE writeForbiddenNorms FOR List[romas::Norm]»
66     «FOREACH this AS item»
67       /*Norm «item.Id»:: «item.Description» */
68          ltl «item.Id» {[]! («item.Activation» && (!«item.Deactivation») && «item.Action»)
69     «ENDFOREACH»
70   «ENDDEFINE »
```

Fig. 7.10 Xpand script: writeForbiddenNorms routine

```
110   «DEFINE writeExecuters FOR List[romas::Executer]»
111     #define max_tasks 5
112   «FOREACH this AS execu»
113   chan «execu.Name»_Pending = [max_tasks] of {mtype, mtype}
114     mtype «execu.Name»_task;
115     mtype «execu.Name»_param1;
116     mtype «execu.Name»_param2;
117
118   /*Signatory party «execu.Name»::«execu.Description»*/
119   proctype «execu.Name»(){
120     end:
121     do
122         :: «execu.Name»_Pending ?? «execu.Name»_normId , «execu.Name»_Action
123     od;
124   }/*end «execu.Name»*/
125   «ENDFOREACH»      «ENDDEFINE »
```

Fig. 7.11 Xpand script: writeExecuters

```
23   proctype Buyer(){
24   end:
25        do
26           :: Buyer_Pending ?? Buyer_task , Buyer_target;
27        od;
28   }/*end Buyer*/
```

Fig. 7.12 mWater Buyer role in PROMELA

```
78   «DEFINE writeContracts FOR List[romas::ContractTemplate]»
79   mtype ={no_initated, executing, finished, interrumped}
80     «FOREACH this AS cont»
81   mtype «cont.Id»_state=no_initiated;
82   /*ContractTemplate «cont.Id»::«cont.Description»*/
83   proctype «cont.Id»(){
84     «cont.Id»_state=executing;
85     {«LET (List) cont.clause AS normList»
86     do«FOREACH normList AS normaL2»
87     «LET (Norm) normaL2 AS norma »
88
89     «IF norma.Deontic.toString()=="OBLIGED" || norma.Deontic.toString()=="PERMITTED" »
90     ::«IF norma.Activation !=null»«norma.Activation»->«ENDIF»
91     «FOREACH norma.Target AS targ»«targ.Name»_Pending ! «norma.Id» «norma.Action» 0;«ENDFOREACH»
92             «ENDIF»
93     «ENDLET»      «ENDFOREACH»
94     od;«ENDLET»
95     «cont.Id»_state=finished;
96     }unless(«cont.Expiration»);
97   } /*end contract «cont.Id»*/
98   «ENDFOREACH»   «ENDDEFINE»
99
```

Fig. 7.13 Xpand script: writeContracts

The status of a contract is represented as a global variable (Fig. 7.13 line 81–Fig. 7.14 line 31). The expiration condition of a contract is represented as the escape sequence of an *unless* statement which includes all the tasks of the contract. This means that if the expiration condition is satisfied, the contract will interrupt its execution (Fig. 7.13 line 96–Fig. 7.14 line 41). Each obligation and permission clause adds the action of the norm to the channel of the corresponding executer that is simulating the execution (Fig. 7.13 lines from 89 to 94–Fig. 7.14 lines 37 and 38).

– *writeInit routine* (Fig. 7.15): This routine creates the *Init process* which is the first process that is executed in a PROMELA code. This process launches the executers processes and the contract processes when they are activated.

```
31  mtype BuyWaterRightContract_state=no_initiated;
32  /*ContractTemplate BuyWaterRightContract::Description*/
33  proctype BuyWaterRightContract(){
34          BuyWaterRightContract_state=executing;
35      {
36      do
37      ::true->Seller_Pending ! norm1 give_water 0;
38      ::true->Buyer_Pending ! norm2 plow 0;
39      od;
40      BuyWaterRightContract_state=finished;
41      }unless(false);
42  } /*end contract BuyWaterRightContract*/
```

Fig. 7.14 mWater BuyWaterRightContrat in PROMELA

```
141 «DEFINE writeInit (List[ContractTemplate] contractList, List[Executer] executerList) FOR romas_model »
142  init(
143 «FOREACH executerList AS execu»
144      run  «execu.Name»();
145 «ENDFOREACH»
146 «IF !contractList.isEmpty»
147      do
148      «FOREACH contractList AS contr»
149      :: («IF contr.Activation !=null» «contr.Activation» && «ENDIF»«contr.Id»_state==no_initiated) ->
150 run «contr.Id»();
151      «ENDFOREACH»
152      od;
153 «ENDIF»
154    )
155    «ENDDEFINE »
```

Fig. 7.15 Xpand script: Init process

In our case study, after generating the PROMELA code, the SPIN model checker was executed from the modeling tool to verify that these contracts and norms were coherent. The SPIN verification shows that the LTL formula *Org2 mWater norm* had been violated. This means that there is an organizational norm or a contract clause that is incoherent with the *Org2 mWater norm*. In this case, the conflict is produced by the clause *PlowService* from the *BuyingWater* contract.

The *Org2 mWater norm* specifies that agents playing the *Buyer* role cannot provide other services, whereas the clause in the contract specifies that the *Buyer* must provide the service *Plow* to the *Seller*.

Therefore, the designer should revise the design of the system and decide if this norm is too strict or if such a contract cannot be performed inside the system. After the redesign, the verifier module could be executed again.

7.5.4 Contributions and Limitations

In this section, we have shown how model checking techniques have been integrated in the ROMAS CASE tool to verify the coherence of the normative context of the system.

Model checking allows a full verification of distributed systems, however, it has serious scalability problems. The reason is that model checking expands all the possible states of the system and verify that the property evaluated is not violated in any of these states.

The verification module was successfully used to verify the coherence of the normative context of the case studies presented in Chap. 9. In order to analyze the scalability of our proposal for verifying the normative context of the system, we have performed two scalability tests.

As is described in this section, the verifiable models generated by the ROMAS tool generate: (1) One LTL property for each prohibition norm that is in the normative context to be evaluated. (2) One process for each role or agent. This process will simulate the execution of each task that was assigned to it. (3) One process for each contract. This process is responsible for sending the role processes a task for each obligatory or permitted action that this contract contains. (4) An init process that initializes the system. After few tests, we found that the most critical variable for the scalability of the verification module is the number of obligation and permitted norms, i.e., the number of actions whose execution the system has to simulate.

In our first scenario, we analyze a system with only one prohibition norm (one LTL property), one contract, and one obliged action per role. The number of roles are modified from 1 to 10, therefore the total number of actions increase also from 1 to 10. As Fig. 7.16 shows that the maximum depth of the states expanded to evaluate the design increases proportionally to the number of actions of the system. In the same way, the verification time and the memory needed to perform the evaluation also increase proportionally to the number of actions of the system.

In our second scenario, we analyze a system with only one prohibition norm (one LTL property), two roles, and a variable number of contracts. The number of contracts varies from 1 to 6. Each contract assigns one action to one role, in that sense, each role executes as many actions as contracts divided by 2. Although in both scenarios the total number of actions increase, the main difference is that in the first scenario, the number of actions per role is fixed to one, whereas in the second scenario the number of roles is fixed but the number of actions that each role executes is variable.

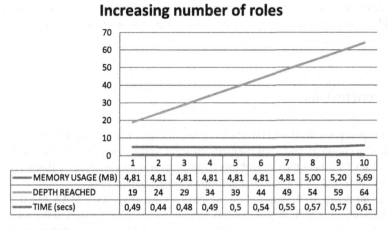

Increasing number of roles

	1	2	3	4	5	6	7	8	9	10
MEMORY USAGE (MB)	4,81	4,81	4,81	4,81	4,81	4,81	4,81	5,00	5,20	5,69
DEPTH REACHED	19	24	29	34	39	44	49	54	59	64
TIME (secs)	0,49	0,44	0,48	0,49	0,5	0,54	0,55	0,57	0,57	0,61

Fig. 7.16 Scalability test 1

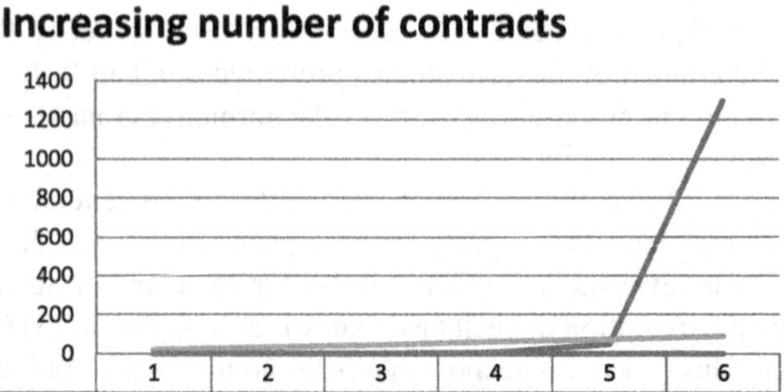

Increasing number of contracts

	1	2	3	4	5	6
▬ MEMORY USAGE (MB)	4,81	4,81	4,90	6,86	49,34	1297,60
▬ DEPTH REACHED	20	34	48	62	76	90
▬ TIME (secs)	0,44	0,5	0,61	0,72	0,81	1,2

Fig. 7.17 Scalability test 2

As Fig. 7.17 shows, in this case the maximum depth of the generated graph and the memory usage increase faster than in the previous case. The verification of a system with six contracts, i.e., a system where two roles execute three actions, each need a huge quantity of memory.

In light of the results of the scalability tests, it is obvious that techniques for improving the scalability of our verification module should be investigated and implemented. This is ongoing work and there are a lot of open issues on the verification of ROMAS models such as the verification of the coherence of the individual and global objectives, between the commitments of the entities of the system and their capabilities, and so on.

Since the ROMAS CASE tool has been implemented as a set of Eclipse plug-ins, it would be easy to add new plug-ins that implement these verifications. The architecture of these new verification modules could be similar to the verification module presented in the previous section. First, a set of Xpand transformation patterns would be defined and then the results of these patterns verified using the integrated SPIN model checker.

7.6 Conclusions

This chapter presents the ROMAS development framework. It is a CASE tool based on MDD technology and implemented as a set of Eclipse plug-ins. The use of Eclipse technology facilitates the extensibility of the system and its interoperability with other Eclipse tools and with any tool that follows the ecore standard.

The modeling tool supports the design of normative open MAS based on the ROMAS metamodel. It means that this tool allows to explicitly design the social and normative context of a system by means of norms and contracts. The tool offers one graphical editor for each view of the ROMAS metamodel (organizational, internal, contract template and activity view). This fact allows representing the model in different diagrams facilitating the design of the system and increasing the clarity of the models.

The development framework also allows the verification of the coherence of the normative context of the system by means of model checking techniques. Verifying the coherence of the normative context is an important topic in the development of normative open systems. Due to the fact that these systems commonly integrate several legal documents and specific regulations, there is a high risk of conflict in the specification of the normative context. Detecting and solving these conflicts at design time produces more trustworthy systems and may avoid critical situations and implementation of uncorrect systems.

The integration of the design and verification of the models in the same tool facilitates the modeling task. Besides, this tool is prepared to integrate different code generation modules in order to generate platform execution code from the verified model. At the moment, there is ongoing work that is generating a code generation module from ROMAS models to code for the Thomas platform.

This development framework has been successfully used to design and verify several case studies [57, 60, 61]. However, as described in Sects. 7.4.3 and 7.5.4, the ROMAS development framework is ongoing work that still have some open issues.

Part III
Evaluation and Case Studies

Chapter 8
ROMAS Approach Evaluation

This chapter analyzes to what extent the ROMAS approach supports the development of normative open MAS.

The rest of the chapter is organized as follows: Sect. 8.1 analyzes how ROMAS supports the development of normative open systems by means of revisiting the criteria presented in Sect. 3.2. Chapter 9 presents some of the case studies modeled using ROMAS and analyzes the lessons learned from the application of this methodology. Section 9.4 summarizes the main conclusions of the ROMAS evaluation.

8.1 ROMAS for Developing Normative Open MAS

This section revises the requirements for developing normative open MAS presented in Chap. 2 in order to analyze to what extent ROMAS supports the development of systems of this kind. This analysis is based on the criteria for evaluating the support of the development of normative open MAS presented in Sect. 3.2. The evaluation criteria are divided in three categories as presented in Chap. 2.

The rest of this section explains the results obtained when evaluation of ROMAS following these criteria. Table 8.1 summarizes the results of the ROMAS evaluation.

8.1.1 Design Abstractions

ROMAS methodology is focused on the analysis and design processes for developing organizational multi-agent systems where agents interact by means of services and where social and contractual relationships are formalized using norms and contracts. Table 8.1 summarizes which design abstractions for developing normative open MAS the ROMAS approach integrates into its architecture, metamodel, and tools. More details about how the ROMAS integrate these design abstractions can be consulted in Chap. 6 and a summary is presented below.

© Springer International Publishing Switzerland 2015   121
E. Garcia et al., *Regulated Open Multi-Agent Systems (ROMAS)*,
DOI 10.1007/978-3-319-11572-6_8

Table 8.1 ROMAS' evaluation

Design abstractions									
Organizations	Services	Normative contexts	Institutional norms	Role norms	Agent norms	Structural norms	Social relationship contracts	Play role contracts	Contractual agreements
Supported	Supported	Supported	Supported	Supported	Supported	Supported	Supported	Supported	Supported
Support during the development process									
Coverage of the lifecycle	Social structure	Requirement norms	Legal documents	System design	Structure considers norms	Contractual agreements	Contract protocols		
Analysis and design	Supported (by integrating Gormas guidelines)	Supported	Supported	Supported	Supported	Partially supported	Not supported		
Evaluation of the final design									
Modeling tool	Code generation	Validation of the requirements	Verification of inconsistences	Tests	Coherence of the normative context	Traceability of the normative context			
Provided	Partially supported	Not supported	Not supported	Not supported	Partially supported	Supported			

In ROMAS, organizations represent a set of individuals and institutions that need to coordinate resources and services across institutional boundaries. Organizations can represent real-life institutions and simulate their functionality and structure. On the other hand, organizations can also be used as an abstraction that represents a set of entities that have common properties, objectives, or regulations.

ROMAS architecture is based on a service-oriented paradigm. The specification of interchanges of functionality and products by means of services at design time allows the designer to focus on what every entity should provide and not on how it is going to be provided. ROMAS also allows to detail how each entity implements the services that provides by means of the specification of tasks and messages. Since the profile of the service is specified separately from the process, different possible implementations can be designed for a service.

ROMAS includes into its metamodel and development process the identification and specification of role, agent, structural and institutional norms. Norms in ROMAS are used to restrict the behavior of the entities of the system. Role norms indicate the norms that any entity playing that role should fulfill. Agent norms are the norms that are associated to a specific entity because of its individual design requirements. Structural norms formalize the structure of the system by means of norms. This fact allow entities to reason and modify the structure of the system at design time. Institutional norms indicate the norms that any entity that is member of an organization should fulfill. Since organizations allow specifying different normative contexts, some norms that are active in one organization or context could not be valid in other contexts.

ROMAS also includes the specification of social relationship contracts, play role contracts, and contractual agreements. Contracts are used to formalize the relationships between entities. These contracts are completely specified by means of the contract template view of the ROMAS metamodel.

8.1.2 Support During the Development Process

ROMAS methodology is structured in five phases that covers the analysis and design of normative open MAS. This is not a linear process but an iterative one, in which the identification of a new element of functionality implies the revision of all the diagrams of the model and the work products produced, so it requires to go back to the appropriate phase. ROMAS specifies a sequences of tasks that should be performed in order to analyze and design the system. These sequences of task are supported by a set of guidelines. The results of these tasks are formalized by means of instances of the ROMAS metamodel.

The first phase, called system specification, includes a set of guidelines for analyzing the system requirements, the goals of the system, and the use cases. It also provides a guideline that helps developers to check the suitability of the ROMAS methodology regarding the requirements of the system to be developed.

The second phase, called organization specification, includes guidelines for analyzing the social structure of the system. First, ROMAS offers a guideline for identifying the roles of the system. Second, in order to define what is the best way to structure these roles, the GORMAS guideline is integrated [9]. The integration of this guideline is possible because both metamodels share the same concepts of role, organization, and agent. Also, both methodologies have been specified using the FIPA standard Design Process Documentation Template, so the social structure guideline can be directly reused in ROMAS.

The third phase, called normative context specification, includes a set of guidelines for identifying and formalizing the normative context of the system, i.e., the norms and contracts that regulate the behavior of the system. Three guidelines are included: (1) Organizational norms guideline that specifies a step-by-step process to identify and formalize restrictions on the behavior of entities gained from the analysis of system requirements. (2) Normative document guideline that specifies a step-by-step process to analyze normative documents in order to identify which restrictions must be implemented in the system. (3) Social contracts guideline that specifies a step-by-step process to identify and formalize social contracts inside a specific organization regarding the role's specification and the structure of the organization. ROMAS also integrate the verification of the normative context into the development process.

The fourth phase, called activity specification, guides developers during the specification of each task, service, and protocol by means of instances of the activity model view of the ROMAS metamodel. However, ROMAS does not provide guidelines that automatize and facilitates the selection of the most suitable implementation for a specific task, service, or protocol. This is an open issue in ROMAS.

The fifth phase, called agent specification, includes a set of guidelines for analyzing the specific requirements of each agent and for selecting which roles should be played in order to achieve their objectives.

8.1.3 Evaluation of the Final Design

ROMAS methodology is supported by a CASE tool called ROMAS tool. As is described in Chap. 7, this case tool allows modeling normative open MAS following the ROMAS metamodel. Using the Eclipse modeling technology, this case tool provides an automatic code generation plug-in that transforms ROMAS models into executable code for the Thomas platform. However, only skeletons of agents and organizations are generated. At the moment, the normative context of the system is not translated. The development of this plug-in is an ongoing work that is out of the scope of this book.

ROMAS integrates in its development process the evaluation of the coherence of the normative context. This guideline is integrated in the case tool by using the Eclipse modeling technology and the Spin model checker (see Sect. 7.5). However, the ROMAS tool does not offer any tool to validate the requirements and verify that there is no inconsistencies between the individual behavior of each entity and the global behavior of the system.

ROMAS does not support the creation of simplified systems prototypes to simulate the behavior of the system. This property would be very useful to experimentally verify the designs and to simulate systems.

ROMAS methodology offers traceability between the requirements of the system and the norms and contracts that form the normative context of the system. Norms and contracts are identified and formalized by means of guidelines that show the specific path that has been followed in order to formalize them. Moreover, the norms identifier attribute is defined following a standard process that allows to trace their origin. The traceability property is very useful to avoid reimplementing the whole system when part of the specification of the system or its normative environment change.

8.2 Comparison with Other Agent Methodologies

The analysis of ROMAS using the criteria presented in Sect. 3.2 is used to compare ROMAS with the methodologies studied in Chap. 2. Tables 8.2 and 8.3 summarize the results of this comparison.[1]

The main difference between ROMAS and O-Mase is that O-Mase does not include in its metamodel the concept of contracts and only few types of norms are considered. As is described in Chap. 2, the use of contract templates during the analysis and design phases allows a complete specification of the legal environment of the system and the relationships between entities without compromising how these entities will implement their commitments. Entities can know what to expect from the other entities.

Gormas methodology shares the concepts of organizations, roles, agents, and norms with ROMAS, however, Gormas does not include the concept of contract in its metamodel.

The initial version of Tropos does not support norms nor contracts. Although Telang et al. [109] enhances Tropos with commitments, nevertheless social contracts are not supported and the proposed development process does not guide developers in the identification or formalization of these contracts.

OperA metamodel differs from ROMAS in the definition of organization. In OperA, organizations are defined as *institutions* and the activity inside these institutions is regulated by means of scenes. However, the semantic meaning and applicability are quite similar to the ROMAS concepts. Both metamodels integrate the use of norms, social contracts, and contractual agreements. The main contribution of ROMAS versus OperA is the integration of guidelines for identifying and formalizing the normative context of a system and the integration of the verification of the coherence of the design during the development process. The lack of these kinds of guidelines is a common issue in all the studied methodologies.

---

[1] The content of these tables is the union of the Tables 3.2, 3.4, 3.6 and 8.1. So, Tables 8.2 and 8.3 do not add new information but they are introduced for clarity reasons.

Table 8.2 ROMAS comparison I

	OMASE	OPERA	TROPOS	GORMAS	ROMAS
Design abstractions					
Organizations	Supported	Supported	Partially supported	Supported	Supported
Services	Supported	Supported	Not supported	Supported	Supported
Normative contexts	Supported	Supported	Not supported	Supported	Supported
Institutional norms	Supported	Supported	Not supported	Supported	Supported
Role norms	Supported	Supported	Not supported	Supported	Supported
Agent norms	Supported	Not supported	Not supported	Supported	Supported
Structural norms	Not supported	Supported	Not supported	Supported	Supported
Social relationship contracts	Not supported	Supported	Partially supported	Not supported	Supported
Play role contracts	Not supported	Supported	Partially supported	Not supported	Supported
Contractual agreements	Not supported	Supported	Partially supported	Not supported	Supported
Support during the development process					
Social structure	Provided	Provided	Not provided	Provided	Provided (by integrating GORMAS guideline)
Requirement norms	Partially provided	Partially provided	Not provided	Partially provided	Provided
Legal documents	Not provided	Not provided	Not provided	Not provided	Provided
System design	Considered	Considered	Not considered	Considered	Considered
Structure considers norms	Part of the normative system is analyzed before but it is not integrated in the guideline.	Part of the normative system is analyzed before but it is not integrated in the guideline.	Not considered	Supported	Supported
Contractual agreements	Not provided	Partially provided	Not provided	Not provided	Partially provided
Contract protocols	Not provided	Partially. It offers a library of patterns for interaction protocols.	Not provided	Not provided	Not provided

Table 8.3 ROMAS comparison II

	OMASE	OPERA	TROPOS	GORMAS	ROMAS
Evaluation of the final design					
Modeling tool	Provided	Provided	Partially provided. The tool does not support norms and contracts	Provided	Provided
Code generation	Partially provided	Partially provided	Not provided	Partially provided	Partially provided
Validation of the requirements	Not supported	Not supported	Partially supported	Not supported	Not supported
Verification of inconsistencies	Not supported	Not supported	Not supported	Not supported	Not supported
Tests	Not supported	Not supported	Not supported	Not supported	Not supported
Coherence of the normative context	Partial verification in the case tool	Partial verification in the case tool	Not supported	Not supported	Partially supported
Traceability of the normative context	Not supported	Not supported	Not supported	Not supported	Supported

Comma [108, 111] and Amoeba [36] are business process methodologies. Although these methodologies can be used to model normative systems, they do not support the design of certain properties that ROMAS supports.

Comma is a methodology for developing cross-organizational business models. This methodology begins from an informally described real-life cross-organizational scenario and produces formal business and operational models. This methodology shares some concepts with ROMAS such the concept of agent, role, goal, task, and commitment. However, the purpose and the level of abstraction of ROMAS and Comma are different. ROMAS is focused on the development of the system by means of analyzing the global purpose of the system and the individual objectives of each agent. Whereas Comma is focused on the specification of business models in terms of agents, roles, goals, tasks, and commitments. Comma does not specify the features of each individual agent. Comma does not support the specification of different organizations, and the social structure of the system is defined only by the commitments between the entities.

Amoeba is a process modeling methodology that is based on commitment protocols similarly to Comma. Amoeba is focused on the specification of business protocols by means of roles, commitments, and low-level interaction protocols. The Amoeba approach is similar to Comma, but Comma lies at a higher level of abstraction containing business goals, tasks, and commitments. Amoeba and ROMAS share the same concept of commitment (called contracts in ROMAS) and roles, however, ROMAS analyzes the system from a higher level of abstraction and also details the social structure and the features of each individual entity.

Chapter 9
Case Studies

This chapter analyzes the usability and benefits of using the ROMAS methodology by means of the analysis of the results when developing different case studies with ROMAS. The selected case studies from different application domains allow analyzing the ROMAS methodology from different dimensions.

Following these case studies are introduced and the lessons learned during their development are discussed.

9.1 CMS Case Study

The conference management system (CMS) is a system to support the management of scientific conferences which involves several aspects from the main organization issues to paper submission and peer review, which are typically performed by a number of people distributed all over the world. The analysis and design of this case study has been detailed in Chap. 6 where it has been used as a running example to exemplify the different phases of the ROMAS methodology.

9.1.1 Lessons Learned and Benefits of Applying ROMAS

This is a common case study in MAS to analyze and exemplify new metamodel proposals and development process approaches [35, 122].

DeLoach et al. [35] present the design of this case study using three different methodologies (O-Mase, Tropos, and Prometheus). The social structures of this case study designed after applying these methodologies are quite similar between them. Comparing these designs with the design obtained by ROMAS, we can conclude that there are several similarities. For example, the roles or the system are mainly the same. The main differences are due to the fact that in the ROMAS analysis we assume that

© Springer International Publishing Switzerland 2015   129
E. Garcia et al., *Regulated Open Multi-Agent Systems (ROMAS)*,
DOI 10.1007/978-3-319-11572-6_9

different and independent conferences that may have its own requirements, features, and legal environment can be integrated in the system.

As is shown in [35], designing the CMS case study with a non-normative approach is possible. However, the CMS system has a complex normative context derived from several legal documents and internal legislations of the institutions participating in the system that bound the behavior of the entities of the system. A design that does not analyze this normative context will rely on the expertise of the developers to include these restrictions on the final implementation. Also, the explicit representation of the norms of the system facilitates the communication with the domain expert and the verification of the correctness of the design.

In ROMAS, a set of guidelines guides developers when identifying and formalizing the normative context of each entity of the system. Developing such a system without a complete development process and without guidelines that help the designers to identify and formalize the normative contexts of the system would require a lot of expertise of the designer. Even for an expert designer it would be easy to miss key design constraints that could be critical for the system.

The use of contracts in the design of the CMS case study creates a flexible system in a regulated context. Developers know exactly what is the expected behavior of every entity and that, as long as they follow the norms of these contracts their implementation will be able to be integrated in the system.

9.2 mWater Virtual Market

Garrido et al. [63, 64] present a case study, called *mWater*, that can be used as a test bed for agreement technologies. The mWater system is an institutional, decentralized framework where users with water rights are allowed to voluntarily trade their rights with other users, in exchange for some compensation, economic or otherwise, but always fulfilling some preestablished rules. In this virtual market based environment, different autonomous entities, representing individuals, groups of irrigators, industries, or other water users, get in contact in order to buy and sell water rights. They are able to negotiate the terms and conditions of the transfer agreement following normative laws. The control of the water supply is distributed by means of water basin institutions, where each basin institution controls the transfer of water rights within their basin. In order to perform an interbasin transfer, the agreement should be authorized by the government of the country.

9.2.1 Applying ROMAS

In this section, a brief overview of the results obtained when applying the ROMAS methodology to the mWater case study is presented. Further details about how the ROMAS methodology has been applied to this case study can be consulted in [59].

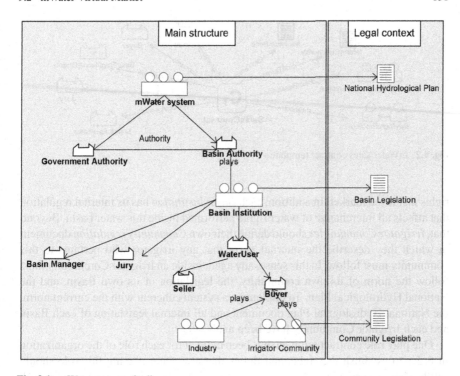

Fig. 9.1 mWater case study diagram

Figure 9.1 presents the global structure of the mWater system using the ROMAS metamodel diagram. The system is composed of two main roles: the *Government Authority* role is the maximum authority in the system, and is responsible for controlling the interchange of water rights among basins, and maintaining the *InterBasins Contracts database*; the *Basin Authority* role represents the highest authority of each basin institution, and is not played by an individual entity but by an institution.

Every *Basin Authority institution* has the following roles: *Water user* role that represents the individual entities that are registered in the system as *Buyers* or *Sellers* to negotiate and interchange water rights. Figure 9.1 specifies that the role *Buyer* can be played by *Irrigator communities* or *Industries*, i.e. this role can be played by individuals or by organizations as a whole. The *Jury* role is responsible for solving disputes between *Water users*. The *Basin Manager* role is responsible for registering the agreements in the *Basin Contracts database*, for ensuring the stability of the market by controlling the fulfillment of contracts and norms, and for maintaining the *Right Holders database*, updated with the information of the status of each water right.

Three types of normative documents are attached to the system Fig. 9.1. The whole system follows the governmental norm specified in the *National Hydrological Plan* that should be fulfilled by any entity and institution that wants to negotiate water

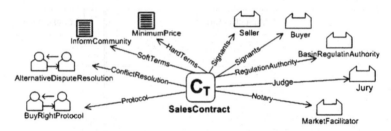

Fig. 9.2 mWater sales contract template

rights inside the market. In addition, each *Basin Institution* has its internal regulation that affects all interchange of water rights performed inside this water basin. Beyond that, *Irrigator Communities* should define their own *Community regulation* document in which they describe the internal norm that any irrigator who pertains to this community must follow. In this sense, any agent inside an Irrigator Community must follow the norm of its own community, the legislation of its own Basin, and the National Hydrological Plan. To implement a system coherent with the current norm, the National Hydrological Plan document and all internal legislation of each Basin and each Irrigator Community have been analyzed.

One play role contract template has been defined for each role of the organization in order to establish the rights and duties that any agent playing this role should fulfill. Therefore,six play role contract templates have been formalized: one for each role of the main organization (*Basin Authority* and *Governmental Authority*), and one for each role described inside the Basin Institution organization.

Following ROMAS, one social relationship contract template should be defined for each pair of roles that must interchange services and products as part of the social structure of the organization. Contracts of this kind should be negotiated and signed by the related entities and not by the organization as whole. However, as is specified in ROMAS, if the terms of the contract are not negotiated by the entities, and the relationship between these agents is determined by their organization, it is not necessary to create a social relationship contract. Instead, the rights and duties of each role over the other are included in their respective play role contracts. In the mWater case study, there is an authority relationship between the *Government Authority* role and the *Basin Authority* role. The terms of this relationship are specified by the main organization of the system regarding the current legislation. Therefore, the rights and duties from one entity to the other are formalized in their respective play role contract and no social relationship contract is created.

The mWater organization establishes some restrictions that every transaction in this market should follow. These restrictions are specified in the sales contract template presented in Fig. 9.2. In this sense, any contractual agreement performed inside the system will fulfill the restrictions established in this contract template.

The model shows that only agents who play the role of *Seller* and *Buyer* can participate in such a contract. Designers recommend the *BuyRight Protocol* in order to define completely this type of contracts. The agent who plays the role of *MarketFacilitator* is the responsible for verifying the correctness and coherence of the final contract. The agent who plays the role of *Jury* will mediate and judge if there is any complaint related to this contract. In order to solve the conflict, the protocol *Alternative Dispute Resolution* will be executed. Moreover, the contract template specifies that the agent who plays the role of *BasinRegulatinAuthority* will be the main authority in the context of this contract.

Every Sale Contract should include the norm *MinimumPrice* = (FORBIDDEN, price <0.07 eur/l), which means that it is forbidden to sell water for less than 0.07 eur/l. The norm *InformCommunity* = (OBLIGATION, Seller, inform(mWaterOrganization)), means that the Seller should inform the community about the final terms and conditions of the agreement. It is a *Soft Term*, i.e. it is only a recommendation. Therefore, during the negotiation process the signatories will decide if this norm should be included or not in the final contract.

9.2.1.1   Lessons Learned and Benefits of Applying ROMAS

The mWater case study is an open virtual market that shares many characteristics and development challenges with any other type of virtual market. Therefore, the lessons learned during the analysis and design of this system could be extensible to other virtual markets' case studies.

One of this challenges is the development of an open environment where entities can be integrated at runtime. ROMAS deals with this challenge by means of standard web services and the specification of social contracts that clearly define under which terms an agent can acquire a role in the system.

Another challenge is how to regulate the interchanges of services or products. The specification of contractual contracts has been very useful to explicitly represent restrictions on the contractual agreements that the entities of the system can perform. The application of the ROMAS methodology to the analysis and design of the mWater system has created a complete design where the normative context of this system has been explicitly formalized.

The development of the mWater case study with other methodology would imply the following issues:

- If the methodology wouldn't specify norms, the restrictions on the behavior of the entities of the system should have been internally specified in the implementation of each entity. Therefore, it would not be secure the integration of entities that had been implemented outside the scope of the system. *Basin institutions* software should have been revised and reimplemented before letting them to be integrated in the system. Moreover, any change in the norms of the system (e.g. a new legislation of the National Hydrological Plan) would require to stop the system and reimplement it before restarting it again.
- If the methodology wouldn't specify contracts, the *water users*, *basin institutions*, and all the entities of the system should know in advance which are the rights, duties, and restrictions that they acquire when entering into the system.

They wouldn't be able to negotiate specific conditions for each entity. Moreover, the relationships between entities wouldn't be formalized. This fact implies that there would not be negotiations of the terms and that no specific norm could be attached to any specific interaction.

- mWater system, as well as many other normative systems, has a complex normative context derived from several legal documents and internal legislations of the institutions participating in the system. Developing such a system without a complete development process and without guidelines that help the designers to identify and formalize the normative contexts of the system would require a lot of expertise of the designer. Even for an expert designer it would be easy to miss a set of norms that could be critical for the system.
- In the case of the mWater system, the verification of the normative context is a fundamental task due to the necessity of verifying that the internal regulation of each *basin institution* is coherent with the Hydrological National Plan.

9.3 The Ceramic Tile Factory System

The manufacturing industry is an interesting domain for applying multi-agent technology, because on the one hand the high development level achieved by this technology allows to tackle with complex problems fields, and on the other hand these systems require software applications that need to be inherently distributed, robust and capable of adapting to the environment. Current manufacturing enterprises should be flexible, responsive, adaptive, and able to cope with the variability of demand. Decisions need to be made fast, be formalized, deal with vast amounts of data, fit business objectives, be right, etc.

The aim of the *ceramic tile factory system* case study is to deal with a real production programming problem in a ceramic tile factory [62]. This problem is considered as one of the main critical issues in a ceramic tile company. It has normally been modeled trying to maximize the simplification of the environment conditions. Nevertheless, the related environment is in fact very dynamic and it reflects the dynamic conditions and constant changes of the ceramic tile sector, such as new client requirements, dynamical work entrance, the availability of machines due to breakdown, etc.

The software application to deal with this problem should be integrated in the ceramic tile and interact with other departments of the factory. The main objectives of this software application are: (1) Automatize the management of raw materials; (2) Offer runtime information about the status of the production; (3) Simulate the execution of a Master Plan in order to help the Commercial department and the Purchases and supplies departments to analyze the results of the execution of a plan; (4) Automatize, monitorize, and manage the manufacturing process; (5) Offer a system that helps coordinating different ceramic tile companies; (6) Schedule the tasks in order to achieve the commitments of the Master Plan; (7) Reschedule the tasks when there is an event that invalidates the previous schedule. The causes for a rescheduling can be a breakdown of a machine, the specification of a new Master Plan due to clients requirements, and so on.

This section briefly summarizes the results of applying the ROMAS methodology to this case study and analyzes the lessons learned and the suitability of ROMAS for developing this case study.

9.3.1 Applying ROMAS

During the first phase of the ROMAS methodology, the requirements of the system are analyzed and the suitability of the ROMAS methodology for the specific case study is studied. The application of the guideline for analyzing the suitability of the ROMAS methodology (see Table 6.8) shows the following results:

- *Distribution*: We are dealing with a distributed system where the information is spread in different data bases that can be in different locations.
- *Intelligent systems*: The system is composed of intelligent systems that can be heterogeneous, proactive and that need to dynamically adapt their process and behavior to handle changes in its requirements.
- *Social structure*: The system does not interact with external entities or institutions. The system is departmentalized and there are authority relationships between its entities.
- *Interoperability*: All the entities of the system are implemented by the same company for a specific tile factory, so the interoperability is not an issue.
- *Regulation*: The system do not have any legal or normative document associated. However, there are many dynamic restrictions derived from the commitments specified in the Master Plan and the features of each machine.
- *Trustworthiness*: The specification of the system using a contract-based approach is not recommendable due to the following reasons: (1) Although the system interact with other departments of the company by providing information about the status of the production, the system is not open to external entities that could play a role in the system; (2) All the entities of the system are implemented by the same company and all of them are under the same sphere of control; (3) The relationships between the entities cannot be negotiated at runtime.

In the light of the results of the requirements of the system and the results of the ROMAS suitability guideline, we can conclude that the best option for developing this case study is to use a methodology that deals with distributed, heterogeneous, and intelligent entities in a social environment. Therefore, the guideline concludes that the ROMAS methodology could be used but it is not the best option and the use of a simpler methodology is recommended.

Despite this results, we decided to continue designing the system following the ROMAS methodology in order to evaluate how the methodology respond to these kinds of systems.

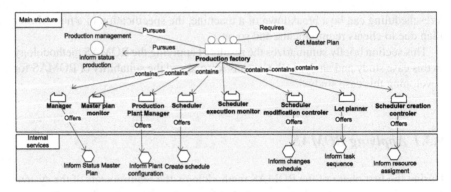

Fig. 9.3 Ceramic tile production organizational view diagram

Figure 9.3 shows the organizational view diagram of this case study. The system is composed of the following roles: (i) *Manager*, that is responsible for the agent organization, maintains integrity between all agents in charge of defining and controlling the schedule, and regulates the cooperation among the different roles; (ii) *Production Plant Manager*, that maintains information about actual plant configuration and knows all restrictions and features of each machine and plant element; (iii) *Scheduler*, that has the ability to schedule tasks and resources; (iv) *Schedule Execution Monitor*, that supervises actual execution of a schedule in a specific plant; (v) *Master Plan Monitor*, that controls possible changes in the Master Plan (according to schedule execution, modification, and creation errors) and informs the Manager role when it identifies an alteration that must be propagated to the Master Plan Generator Process; (vi) *Schedule Modification Controller*, that maintains information about changes needed for adjusting the schedule because of failures in the manufacture process; (vii) *Lot Planner*, that manages all information about the task sequence; (viii) Schedule Creation Controller, that oversees the information about a new schedule order, more specifically about resource assignment for a specific Master Plan Lot.

The analysis of the normative context of the system identifies the norms to specify who can access to the resources and databases of the system and who can access to the services of the system. One play role contract template is specified for each role. Since there is no normative document attached to this case study, the terms of these contract templates are only obligation norms that specify the tasks and services that agents playing these roles should provide. These terms are not negotiable so there is no negotiation protocol attached to these contracts.

9.3.1.1 Lessons Learned and Benefits of Applying ROMAS

Previous works [62, 67] analyze this case study using the Ingenias methodology [95]. Comparing the designs obtained by ROMAS and Ingenias, we can observe that both identify the same roles of the system and a similar social structure. This version of the

Ingenias methodology does not support the specification of services. Although this case study is not an open system in the sense that external entities cannot become part of the system, we consider that the use of services is interesting in order to modularize the system. The specification of activities by means of services allows parts of the system to be modified without affecting the rest of the system.

The conclusion obtained from the experience of designing this case study using the ROMAS methodology is coherent with the recommendation offered by the ROMAS suitability guideline, i.e. it is possible to design a system of this kind with ROMAS but it is not recommendable. The main reasons are that using ROMAS we have generated diagrams that do not offer meaningful information and that we have followed complex guidelines and steps with no result. For example, the play role contract templates are not useful here because there are no external entities trying to play these roles, there is no possibility of negotiating the terms of these contracts, and because they only specify the functionality of these roles. This functionality is also specified in the internal view diagram that specify each role, so this information is redundant.

The identification and formalization of the normative context guidelines and steps of the process are examples of useless steps of the process for this case study. There is no legal document attached to this case study and the relationships between the entities of the system are fixed by the requirements of the system and cannot be negotiated at runtime.

The design offered by ROMAS would be interesting if the case study would include the possibility of distributing the production among different companies. In that case, contracts would be necessary to formalize and negotiate the commitments that each company acquire. It would also be interesting if some of the entities of the system could have been implemented by external entities. However, regarding the initial specification of the case study a non-normative and noncontract-based methodology is more suitable for developing this system.

9.4 Conclusions

ROMAS contributes to the state of the art defining a methodology that guides developers during the analysis and design of normative open MAS. ROMAS methodology deals with some of the open issues detected in Chap. 2:

- ROMAS integrates into its metamodel the most important design abstractions for developing normative open MAS selected in Chap. 2 after the study of the requirements for developing such systems. ROMAS integrates the concepts of agent, role, organization, service, norm, and contract during the whole development process.
- Our experiences with the development of the case studies presented above show the importance of these design abstraction in the analysis and design of normative open systems.

  The autonomous and heterogeneous entities that interact in a normative open system are represented by means of agents. The social structure of the system is specified by means of roles and organizations. Roles group functionality into an abstract concept similar to a job in real-life systems. Organizations allow to structure the system in different modules, create different normative contexts and restrict the communication between different parts of the system, simulate real-life institutions, and group different agents in order to achieve a common objective. Services allow to separate the functionality that is offered or requested by an agent from the final implementation of these functionalities. Services also provide a standard interface that facilitate the implementation of open systems creating interoperable systems. Norms and contracts define the normative context of a system. Norms restrict the behavior of the entities and organizations of the system by specifying which behavior is permitted, obliged, or forbidden. The use of contracts to define the social and contractual relationships between entities allows the system to operate with expectations of the behavior of others, but providing flexibility in how they fulfill their own obligations. The social structure of the system is represented by means of social contracts between roles and the roles are defined by means of play contracts. Therefore, the entities of the system can reason about their social structure at runtime and changes this structure by means of changing the norms described in these contracts.

- ROMAS offers a complete set of guidelines that guides developers from the initial requirement analysis to the definition of concrete tasks and interactions. The whole development process is guided by the global goals of the system and it also takes into account the individual goals of each autonomous entity that interact with the system.

  The main contribution of the ROMAS methodology and what differs it from any other approach is that it integrates into the development process guidelines for identifying and formalizing the normative context of the system. ROMAS also integrates the verification of the normative context of the system in the development process.

  The ROMAS development process and guidelines have successfully guided developers during the analysis and design of the case studies. This methodology has allowed to analyze and formally represent during the design the requirements of those systems and their internal features and regulations.

  Although ROMAS' support during the development process is quite complete, there are still open issues such as the lack of guidelines for selecting the most suitable interaction or negotiation protocol regarding the requirements of the system.

- ROMAS CASE tool supports modeling normative open MAS based on the ROMAS methodology. This CASE tool also provides automatic verification of parts of the normative context by means of model checking techniques. At the moment, the implementation of this CASE tool is an ongoing work. More work is still needed in order to improve the scalability of the verification module, increase the number of properties verified (like the coherence between the individual and global objectives) and integrate simulation facilities.

- The description of the methodology using a FIPA standard method allow the reuse of parts of the ROMAS methodology into other development process, as well as, the use of other methodologies' fragments into ROMAS. It also facilitates the comparison between methodologies. It can reduce the time that a system analyst needs to learn a new methodology.

- Our experiences with the case studies show that even when a system is going to be implemented in an agent platform that does not support norms or contracts the analysis and design of the system by ROMAS is beneficial. Using ROMAS, the system is completely specified and developers can know the expected behavior and restrictions of each entity of the system. In that sense, the implementation of a system can be performed by different providers using different technology because all of them know what to expect from the others. At the moment there are some normative open platforms like Thomas [29], but as far as we know, there is no agent platform that integrates contracts. If the developer would like to use the abstraction of contracts in the implementation, he/she would have to implement the contracts without the platform support.

Chapter 10
Conclusions

The increase of collaborative work, the decentralization of processes, and the interaction of entities and institutions in regulated environments highlight the need of new developing approaches. Multi-agent systems technology has found a suitable approach for dealing with systems of this kind; however, there are still gaps in the support that this technology offers. In this book, we have addressed several of the problems derived from the analysis and design of such systems. The main contributions of this book to the state of the art are:

- A study of the requirements for developing normative open systems and the suitability of developing them using the multi-agent systems paradigm.
- A discussion about how current agent methodologies support the development of normative open systems.
- The specification of a new MAS architecture and metamodel that allow the complete specification of normative open MAS.
- The specification of a development process and a set of guidelines that help the designer during the analysis and design of normative open MAS.
- The design and implementation of a development framework that includes a modeling tool and a prototype module for formally verifying MAS designs.

In Sect. 1.2, we analyze how the properties of the multi-agent systems paradigm can be applied to the development of normative open systems concluding that MAS is a suitable approach for developing these kinds of systems.

In order to perform a deeper analysis about what is necessary for developing normative open systems using a multi-agent approach, Chap. 2 summarizes the most important issues when developing systems of this kind and analyzes to what extent current approaches support the development of these systems. The requirements for developing these systems have been rewritten as an evaluation criteria questionnaire that allows analyzing and comparing different approaches. Considering the general study of the state of the art and the comparison of methodologies presented in the previous section, we conclude that there is no complete methodological approach for analyzing and designing normative systems. The most important open issues where

© Springer International Publishing Switzerland 2015
E. Garcia et al., *Regulated Open Multi-Agent Systems (ROMAS)*,
DOI 10.1007/978-3-319-11572-6_10

related to the lack of guidelines for identifying and formalizing the normative context of the system, and the lack of verification tool to check the coherence, completeness, and validity of the designs obtained through the proposed development process. A full list of the open issues is presented in Sect. 3.3.

Chapter 6 presents ROMAS a new methodology focused on the analysis and design of normative open MAS. This methodology is designed in order to deal with some of the open issues in the development of these kinds of systems. ROMAS is based on a well-defined metamodel that integrates the main concepts of agent, role, organization, service, norm, and contract. It offers an organizational structure where entities interact between them by means of standard services and whose behavior is bounded by norms and contracts. Norms in ROMAS indicate the actions that are permitted, obliged, or forbidden inside the system. Contracts are used to formalize the social and contractual relationships between entities. Social contracts define the social structure of the system as a result of the negotiation between the entities of the system instead of specifying and fixed and inflexible structure. Contractual agreements are formalized by means of contracts at design time in order to specify which kind of relationships are allowed in the system and under which terms.

The ROMAS development process guides designers from the requirements analysis and the formalization of the objectives of the system and the individual entities to the low-level specification of the contract templates that restrict the relationship between the entities, the specification of each individual agent and the specification of the interaction protocols. The whole process is supported by a set of guidelines that help developers with design decision such how to identify the roles of the system and how to identify and formalize the normative context of the system. The verification of the coherence of the system is integrated in the development process.

The ROMAS development process is specified in Chap. 6 using the FIPA standard Design Process Documentation Template. The use of this template ensures the completeness of the specification, facilitates the comparison of ROMAS with other methodologies, reduces the learning time for users used to this standard, and allows to export and import development fragments from/to other methodologies specified with this standard.

The ROMAS methodology is supported by a development framework as presented in Chap. 7. It is a CASE tool based on MDD technology that was implemented as a set of Eclipse plug-ins. The use of Eclipse technology facilitates the extensibility of the system and its interoperability with other Eclipse tools or any tool that follows the ecore standard. This framework is composed of a modeling tool that support the design of normative open MAS based on the ROMAS metamodel, and a verification module based on model checking techniques. The verification module allows the verification of the coherence of the normative context of the system. This development framework has found useful to design and verify several case studies; however, it is an ongoing work that still have open issues as is described in Sects. 7.4.3 and 7.5.4.

The methodology and development framework presented in this book has been evaluated theoretically by means of the study of how this approach fulfills the evaluation criteria questionnaire presented in Chap. 2 and the comparison with other current methodologies. Our proposal was also evaluated empirically by means of its application on different case studies from completely different domains and with a wide range of different features. These evaluations, presented in Chap. 8, show that ROMAS is a suitable approach for the development of normative open systems. ROMAS contributes to the state of the art by offering a completely guided analysis and design that is based on a solid metamodel. ROMAS metamodel allows a high-level abstraction design of systems of this kind by means of the specification of organizations, services, and contracts. This metamodel also allows a low level of abstraction design by means of the specification of individual agents, their internal features, and the detail about their interactions. The development guidelines offered by the ROMAS development process have found very useful when designing large and complex systems.

Next section analyzes some of the limitations of this work and presents new future lines of work.

10.1 Limitations and Future Work

ROMAS is an ongoing active research project. ROMAS contributes to the state of the art in the analysis and design of normative open systems. However, there are still some open issues on this topic that ROMAS does not deal with and that offer potential areas for further work.

- Although Chap. 2 introduces a mechanism for analyzing and comparing agent methodologies, the evaluation and comparison of software methodologies is still an open research topic. In the future, we plan to develop a system that allows performing a deeper comparison and that includes software metrics to quantitatively evaluate software methodologies performance.
- ROMAS development process lacks guidelines for specifying the most suitable interaction protocol regarding a set of requirements and restrictions. We are planning to integrate a guideline that allows reusing interaction patterns.
- ROMAS development framework is still under development. The verification module should be improved in order to solve the scalability problems discussed in Sect. 7.5.4. Besides, it should be extended in order to deal with the semantic verification of the coherence between permissions and obligations of contradictory actions, to deal with the verification of the coherence between the individual and global objectives, and the coherence between the commitments of each entity and its internal features and functionality.
- The analysis and design of systems by means of norms and contracts facilitate the final implementation because the expected behavior of each entity and its interactions with its environment are completely specified. Developers know exactly what should be implemented without compromising how they are going to implement it. We consider that the implementation of normative open systems under a platform that explicitly represent contracts would facilitate the implementation task, would

References

1. Eclipse—an open development platform (2011), http://www.eclipse.org/
2. Eclipse—xpand plug-in (2011), http://www.eclipse.org/modeling/m2t/?project=xpand
3. Meta object facility (mof) 2.0 query/view/transformation specification (2007), ftp://ftp.omg.org/pub/docs/omg/00-11-05.pdf. Accessed 07 July 2007 (Object Management Group, Document ad)
4. M. Amor, L. Fuentes, A. Vallecillo, Bridging the gap between agent-oriented design and implementation using mda, in *International Workshop on Agent-Oriented Software Engineering. LNCS*, ed. by J. Odell, P. Giorgini, J. P. Möller (Springer Berlin Heidelberg, New York, NY, USA, 2004), pp. 93–108
5. J.L. Arcos, M. Esteva, P. Noriega, J.A. Rodríguez-Aguilar, C. Sierra. Engineering open evironments with electronic institutions. Eng. Appl. Artif. Intell. **18**(2), 191–204 2005 (Engineering Applications of Artificial Intelligence Top Cited Article 2005–2010)
6. E. Argente, GORMAS: Guas para el desarrollo de sistemas multiagente abiertos basados en organizaciones. Ph.D. thesis, Departamento de Sistemas Informaticos y Computacion, Universidad Politecnica de Valencia, 2008
7. E. Argente, GORMAS: Guìas para el desarrollo de Sistemas Multiagenteabiertos basados en organizaciones, Ph.D. thesis, Departament de Sistemes Informàtics i Computació, Universitat Politècnica de València, 2008
8. E. Argente, V. Botti, C. Carrascosa, A. Giret, V. Julian, M. Rebollo, An abstract architecture for virtual organizations: the THOMAS approach. Knowl. Inf. Syst. **29**(2), 379–403 (2011)
9. E. Argente, V. Botti, V. Julian, Gormas: an organizational-oriented methodological guideline for open mas, in *Proceedings of the AOSE*, (2009), pp. 85–96
10. E. Argente, V. Botti, V. Julian, Organizational-oriented methodological guidelines for designing virtual organizations, in *Distributed Computing, Artificial Intelligence, Bioinformatics, Soft Computing, and Ambient Assisted Living. LNCS*, vol. 5518, ed. by M.-P. Gleizes, J.J. Gomez-Sanz (Springer Berlin Heidelberg, Budapest, Hungary, 2009), pp. 154–162
11. G. Boella, J. Hulstijn, L.W.N. van der Torre, Virtual organizations as normative multiagent systems. in *HICSS* (IEEE Computer Society, 2005)
12. G. Boella, G. Pigozzi, L. van der Torre, in *Normative Multi-Agent Systems*, ed. by G. Boella, P. Noriega, G. Pigozzi, H. Verhagen. Normative systems in computer science–ten guidelines for normative multiagent systems, (Schloss Dagstuhl–Leibniz–Zentrum fuer Informatik, Germany, 2009)
13. O. Boissier, J. Padget, V. Dignum, G. Lindemann, E. Matson, S. Ossowski, J. Sichman, J. Vazquez-Salceda, Coordination, organizations, institutions and norms in multi-agent systems. in LNCS (LNAI), vol 3913 (Springer, Berlin, 2006), pp. 25–26

© Springer International Publishing Switzerland 2015
E. Garcia et al., *Regulated Open Multi-Agent Systems (ROMAS)*,
DOI 10.1007/978-3-319-11572-6

14. R.H. Bordini, M. Fisher, W. Visser, M. Wooldridge, Verifying multi-agent programs by model checking. in *Autonomous Agents and Multi-Agent Systems*, vol 12 (Kluwer Academic Publishers, USA, 2006), pp. 239–256

15. V. Botti, A. Garrido, A. Giret, P. Noriega, The role of MAS as a decision support tool in a water-rights market. in *Post-proceedings workshops International Conference on Autonomous Agents and MultiAgent Systems*, vol 7068 (Springer, Berlin, 2011), pp. 35–49

16. L. Braubach, A. Pokahr, W. Lamersdorf, A universal criteria catalog for evaluation of heterogeneous agent development artifacts. in *Sixth International Workshop From Agent Theory to Agent Implementation (AT2AI–6)*, 2008

17. T. Breaux, Exercising due diligence in legal requirements acquisition: a tool-supported, frame-based approach. in *Proceedings IEEE International Requirements Engineering Conference*, 2009, pp. 225–230

18. T.D. Breaux, D.L. Baumer, Legally reasonable security requirements: a 10-year ftc retrospective. Comput. Secur. **30**(4), 178–193 (2011)

19. T.D. Breaux, M.W. Vail, A.I. Anton, Towards regulatory compliance: extracting rights and obligations to align requirements with regulations. in *IEEE International Requirements Engineering Conference, RE '06* (IEEE Computer Society, Washington, DC, USA, 2006) pp. 46–55

20. P. Bresciani, A. Perini, P. Giorgini, F. Giunchiglia, J. Mylopoulos, Tropos: an agent-oriented software development methodology. Auton. Agent. Multi-Agent Syst. **8**(3), 203–236 (2004)

21. C. Carabelea, O. Boissier, Coordinating agents in organizations using social commitments. Electron. Notes Theoret. Comput. Sci. **150**(3), 73–91 (2006)

22. H.L. Cardoso, E. Oliveira, A contract model for electronic institutions. Int. Conf. Coord. Organ. Institutions Norms Agent Syst. **III**, 27–40 (2008)

23. A. Castor, R.C. Pinto, C.T.L.L. Silva, J. Castro, Towards requirement traceability in tropos. in *Proceedings of the WER* (2004), pp. 189–200

24. R. Centeno, M. Fagundes, H. Billhardt, S. Ossowski, Supporting medical emergencies by mas. in *Agent and Multi-Agent Systems: Technologies and Applications. LNCS*, vol 5559 (Springer, Heidelberg, 2009), pp. 823–833

25. L. Cernuzzi, G. Rossi, On the evaluation of agent oriented modeling methods. in *Proceedings of the OOPSLA 02—Workshop on Agent-Oriented Methodologies* (2002), pp. 21–30

26. A.K. Chopra, F. Dalpiaz, P. Giorgini, J. Mylopoulos, Modeling and reasoning about service-oriented applications via goals and commitments. in *International Conference on Advanced Information Systems Engineering* (Springer-Verlag, Berlin, 2010), pp. 113–128

27. L. Coutinho, A. Brando, J. Sichman, O. Boissier, Model-driven integration of organizational models, in *Workshop on Agent Oriented Software Engineering*, 2008

28. N. Criado, E. Argente, V. Botti, A normative model for open agent organizations. Int. Conf. Artif. Intell. **1**, 101–107 (2009)

29. N. Criado, E. Argente, V. Botti, THOMAS: an agent platform for supporting normative multi-agent systems. J. Logic Comput. **91**(2) 169–215 2011

30. K.H. Dam, Evaluating and Comparing Agent-Oriented Software Engineering Methodologies. Master's thesis, Master of Applied Science in Information Technology—RMIT University, Australia, 2003

31. R. Darimont, M. Lemoine, Goal-oriented analysis of regulations. in *ReMo2V 06—W. Regulations Modelling and their Validation and Verification—CAISE06*, vol 241 (2007)

32. S. DeLoach, Omacs a framework for adaptive, complex systems. in *Handbook of Research on Multi-AGent Systems: Semantics and Dynamics of Organizational Models* (IGI Global, New York, 2009), pp. 76–104

33. S.A. DeLoach, Developing a multiagent conference management system using the O-mase process framework. in *International Conference on Agent-oriented Software Engineering VIII* (2008) pp. 168–181

34. S.A. DeLoach, J.C. Garcia-Ojeda, O-mase; a customisable approach to designing and building complex, adaptive multi-agent systems. Int. J. Agent-Oriented Softw. Eng. **4**(3), 244–280 (2010)

35. S.A. DeLoach, L. Padgham, A. Perini, A. Susi, J. Thangarajah, Using three aose toolkits to develop a sample design. Int. J. Agent-Oriented Softw. Eng. **3**, 416–476 (2009)
36. N. Desai, A.K. Chopra, M.P. Singh, Amoeba: A methodology for modeling and evolving cross-organizational business processes. ACM Trans. Softw. Eng. Methodol. **19**(2), 6:1–6:45 (2009)
37. I. Dickinson, M. Wooldridge, Agents are not (just) web services: investigating bdi agents and web services. in *The Workshop on Service-Oriented Computing and Agent-Based Engineering at AAMAS*, 2005
38. F. Dignum, V. Dignum, J. Padget, J. Vázquez-Salceda, Organizing web services to develop dynamic, flexible, distributed systems. in *International Conference on Information Integration and Web-based Applications Services* (2009), pp. 225–234
39. F. Dignum, V. Dignum, J. Thangarajah, L. Padgham, M. Winikoff, Open agent systems? *International Workshop on Agent Oriented Software Engineering (AOSE) in AAMAS07*, 2007
40. V. Dignum, A model for organizational interaction:based on agents, founded in logic, Ph.D. thesis, Utrecht University, 2003
41. V. Dignum, F. Dignum, Coordinating tasks in agent organizations. or: Can we ask you to read this paper? COIN@ECAI'06: Workshop on Coordination, Organization, Institutions and Norms in MAS, 2006
42. V. Dignum, F. Dignum, A landscape of agent systems for the real world. Technical report 44-cs-2006-061, Institute of Information and Computing Sciences, Utrecht University, 2006
43. V. Dignum, J.-J. Meyer, F. Dignum, H. Weigand, Formal specification of interaction in agent societies. *Formal Approaches to Agent-Based Systems*, vol. 2699 (Springer, Berlin Heidelberg, 2003), pp. 37–52
44. V. Dignum, J. Vazquez-Salceda, F. Dignum, Omni: introducing social structure, norms and ontologies into agent organizations, in *Programming Multi-Agent Systems.LNCS*, vol. 3346, ed. by R. Bordini, M. Dastani, J. Dix, A. Seghrouchni (Springer, Berlin, 2005), pp. 181–198
45. R. Dorofeeva, K. El-Fakih, S. Maag, A.R. Cavalli, N. Yevtushenko, Fsm-based conformance testing methods: a survey annotated with experimental evaluation. in Information and Software Technology, vol 52 (Butterworth-Heinemann, Newton, 2010), pp. 1286–1297
46. S. Fenech, G.J. Pace, G. Schneider, Automatic conflict detection on contracts. in *International Colloquium on Theoretical Aspects of Computing, ICTAC '09* (2009), pp. 200–214
47. J. Ferber, O. Gutknecht, F. Michel, From Agents to Organizations: an Organizational View of Multi-Agent Systems. in *Agent-Oriented Software Engineering VI*, ed. by P. Giorgini, J. Muller, J. Odell. *LNCS*, vol 2935 (Springer-Verlag, Berlin, 2004), pp. 214–230
48. J. Ferber, F. Michel, J. Bez-Barranco, Agre : integrating environments with organizations. Environ. Multi-agent Syst. **3374**, 48–56 (2005)
49. R.F. Fernández, I.G. Magariño, J.J. Gómez-Sanz, J. Pavón, Integration of web services in an agent oriented methodology. J. Int. Trans. Syst. Sci. Appl. **3**, 145–161 (2007)
50. R.F. Fernandez, I.G. Magarinyo, J.J. Gomez-Sanz, J. Pavon, Integration of web services in an agent oriented methodology. J. Int. Trans. Syst. Sci. Appl. **3**, 145–161 (2007)
51. FIPA, Design process documentation template standard specification (2012), http://fipa.org/specs/fipa00097/index.html
52. I. Garca-Magario, J. Gmez-Sanz, R. Fuentes-Fernndez, Ingenias development assisted with model transformation by-example: A practical case, in *7th International Conference on Practical Applications of Agents and Multi-Agent Systems (PAAMS 2009), Advances in Soft Computing*, vol. 55, ed. by Y. Demazeau, J. Pavn, J. Corchado, J. Bajo (Springer, Berlin, 2009), pp. 40–49
53. E. Garcia, E. Argente, A. Giret, Emfgormas: a case tool for developing service-oriented open mas. in *AAMAS '10: Proceedings of the 9th International Conference on Autonomous Agents and Multiagent Systems* (2010), pp. 1623–1624
54. E. Garcia, E. Argente, A. Giret, V. Botti, Issues for organizational multiagent systems development. in *AT2AI at AAMAS08* (2008), pp. 59–65
55. E. Garcia, A. Giret, V. Botti, On the evaluation of mas development tools. in *International Conference on Artificial Intelligence in Theory and Practice (IFIP AI)*, vol 276 (Springer Boston, 2008), pp. 35–44

56. E. Garcia, A. Giret, V. Botti, Software engineering for service-oriented MAS. *Cooperative Information Agents XII. LNAI*, vol 5180 (Springer, Prague, Czech Republic, 2008), pp. 86–100

57. E. Garcia, A. Giret, V. Botti, A model-driven CASE tool for developing and verifying regulated open MAS. Sci. Comput. Program. **76**(6), 695–704 (2013)

58. E. Garcia, A. Giret, V. Botti, Evaluating software engineering techniques for developing complex systems with multiagent approaches. Inf. Softw. Technol. **53**, 494–506 (2011)

59. E. Garcia, A. Giret, V. Botti, Regulated open multi-agent systems based on contracts Inf. Syst. Dev. **10**, 243–255 (2011)

60. E. Garcia, A. Giret, V. Botti, Developing regulated open multi-agent systems. in International Conference on Agreement Technologies (2012), pp. 12–26

61. E. Garcia, G. Tyson, S. Miles, M. Luck, A. Taweel, T.V. Staa, B. Delaney, An Analysis of Agent-Oriented Engineering of e-Health Systems. in 13th International Workshop on Agent-Oriented Software Engineering (AOSE—AAMAS) (2012), pp. 117–128

62. E. Garcia, S. Valero, E. Argente, A. Giret, V. Julian, A FAST method to achieve flexible production programming systems. IEEE Trans. Syst. Man Cybern. Part C Appl. Rev. **38**(2), 242–252 (2008)

63. A. Garrido, A. Giret, V. Botti, P. Noriega, mWater, a Case Study for Modeling Virtual Markets, in *New Perspectives on Agreement Technologies* (Springer, Heidelberg, 2012)

64. A. Garrido, A. Giret, P. Noriega, mWater: a Sandbox for Agreement Technologies. in *CCIA 2009*, vol 202 (IOS Press, The Netherlands, 2009), pp. 252–261

65. J.M. Gascueña, E. Navarro, A. Fernández-Caballero, Model-driven engineering techniques for the development of multi-agent systems. Eng. Appl. Artif. Intell. **25**(1), 159–173 (2012)

66. B. Gateau, O. Boissier, D. Khadraoui, E. Dubois, Moiseinst: an organizational model for specifying rights and duties of autonomous agents. Environ. Multi-Agent Syst. III **4389**, 41–50 (2007)

67. A. Giret, E. Argente, S. Valero, P. Gmez, V. Julian, Applying Multi Agent System Modelling to the Scheduling Problem in a Ceramic Tile Factory. in *Mass Customization Concepts-Tools-Realization IMCM'05* (GITO-Verlag, Berlin, 2005), pp. 151–162

68. A.S. Giret, ANEMONA: Una Metodologia Multiagente para Sistemas Holonicos de Fabricacion. Ph.D. thesis, Departamento de Sistemas Informaticos y Computacion, Universidad Politecnica de Valencia, 2005

69. B. Gteau, O. Boissier, D. Khadraoui, Multi-agent-based support for electronic contracting in virtual enterprises. IFAC Symp. Inf. Control Probl. Manuf. (INCOM) **150**(3), 73–91 (2006)

70. R. Hermoso, R. Centeno, H. Billhardt, S. Ossowski, Extending virtual organizations to improve trust mechanisms (short paper). in *Proceedings 7th International Conference on Autonomous Agents and Multiagent Systems* (2008), pp. 1489–1472

71. C.D. Hollander, A.S. Wu, The current state of normative agent-based systems. J. Artif. Soc. Soc. Simul. **14**(2), 6 (2011)

72. G. Holzmann, *Spin Model Checker, the: Primer and Reference Manual* (Addison-Wesley Professional, Reading, 2003)

73. G.B. Horlin, R.V. Lesse, A survey of multi-agent organizational paradigms. Knowl. Eng. Rev. **19**(04), 281–316 (2004)

74. B. Horling, Quantitative organizational modeling and design for multi-agent systems, Ph.D. thesis, 2006

75. B. Horling, V. Lesser, A survey of multi-agent organizational paradigms. Knowl. Eng. Rev. **19**(4), 281–316 (2004)

76. F.-S. Hsieh, Automated negotiation based on contract net and petri net, in E-Commerce and Web Technologies. LNCS, vol 3590 (Springer, Berlin, 2005), pp. 148–157

77. M. Huhns, M. Singh, Reseach directions for service-oriented multiagent systems. IEEE Internet Comput Serv.-Oriented Comput. Track. 9(1), 75–81 (2005)

78. D. Isern, D. Snchez, A. Moreno, Organizational structures supported by agent-oriented methodologies. J. Syst. Softw. **84**(2), 169–184 (2011)

79. M. Jakob, M. Pěchouček, S. Miles, M. Luck, Case studies for contract-based systems. in *International Conference on Autonomous Agents and MultiAgent Systems* (2008), pp. 55–62

80. V. Julian, M. Rebollo, E. Argente, V. Botti, C. Carrascosa, A. Giret, *Using THOMAS for Service Oriented Open MAS* (Springer, Berlin, 2009), pp. 56–70
81. G. Kardas, Model-driven development of multi-agent systems: a survey and evaluation. Knowl. Eng. Rev. **28**(04):479–503 (2012) (In Press)
82. M. Kollingbaum, I.J. Jureta, W. Vasconcelos, K. Sycara, Automated requirements-driven definition of norms for the regulation of behavior in multi-agent systems, *AISB, Workshop on Behaviour Regulation in Multi-Agent Systems* ((Aberdeen, UK, Scotland, 2008)
83. T. Kovše, B. Vlaovič, A. Vreže, Z. Brezočnik. Eclipse plug-in for spin and st2msc tools-tool presentation. in *International SPIN Workshop on Model Checking Software* (2009), pp. 143–147
84. C.-E. Lin, K.M. Kavi, F.T. Sheldon, K.M. Daley, R.K. Abercrombie, A methodology to evaluate agent oriented software engineering techniques. in *Hawaii International Conference on System Sciences* (2007), p. 60
85. A. Lomuscio, H. Qu, M. Solanki, Towards verifying contract regulated service composition. Auton. Agents Multi-Agent Syst. **24**(3), 345–373 (2012)
86. M. Luck, L. Barakat, J. Keppens, S. Mahmoud, S. Miles, N. Oren, M. Shaw, A. Taweel, Flexible behaviour regulation in agent based systems. in *Collaborative Agents—Research and Development*. LNCS, vol 6066, ed. by C. Guttmann, F. Dignum, M. Georgeff (Springer, Melbourne Australia, 2011), pp. 99–113
87. F. Meneguzzi, S. Modgil, N. Oren, S. Miles, M. Luck, N. Faci, Applying electronic contracting to the aerospace aftercare domain. Eng. Appl. Artif. Intel. **25**(7), 1471–1487 (2012)
88. J.-J.C. Meyer, R.J. Wieringa (eds.), *Deontic Logic in Computer Science: Normative System Specification* (Wiley, Chichester, 1993)
89. S. Miles, N. Oren, M. Luck, S. Modgil, N. Faci, C. Holt, G. Vickers, Modelling and Administration of Contract-Based Systems. in *Symposium on Behaviour Regulation in Multi-Agent Systems at AISB 2008* (2008)
90. M. Morandini, C.D. Nguyen, A.S.A. Perini, A. Susi, Tool-supported development with tropos: the conference management system case study. in Agent Oriented Software Engineering (AOSE), at AAMAS (2007)
91. D. Okouya, V. Dignum, Operetta: A prototype tool for the design, analysis and development of multi-agent organizations (demo paper). in *AAMAS* (2008), pp. 1667–1678
92. N. Oren, S. Panagiotidi, J. Vázquez-Salceda, S. Modgil, M. Luck, S. Miles, Towards a formalisation of electronic contracting environments. in *International Conference on Coordination, Organizations, Institutions, and Norms in Agent Systems* (2009), pp. 156–171
93. N. Osman, D. Robertson, C. Walton, Run-time model checking of interaction and deontic models for multi-agent systems. in *AAMAS '06: Proceedings of the fifth International Joint Conference on Autonomous Agents and Multiagent Systems* (ACM, New York, NY, USA, 2006), pp. 238–240
94. G. Pace, C. Prisacariu, G. Schneider, Model checking contracts a case study. in *Automated Technology for Verification and Analysis*. LNCS, vol 4762, ed. by K.S. Namjoshi, T. Yoneda, T. Higashino, Y. Okamura (Springer, Tokyo, Japan, 2007), pp. 82–97
95. J. Pavon, J. Gomez-Sanz, R. Fuentes, The ingenias methodology and tools. in Agent-Oriented Methodologies, vol IX (Henderson-Sellers, 2005), pp. 236–276
96. A. Perini, A. Susi, Automating model transformations, in *agent-oriented modelling, in Agent-Oriented Software Engineering VI*. LNCS, vol 3950 ed. by J. Mller, F. Zambonelli (Springer, Berlin, 2006), pp. 167–178
97. M.P. Singh, M.N. Huhns, *Service-Oriented Computing Semantics, Processes* (Agents, Wisley, London, 2005)
98. M. Rodrigo, S. Valero, C. Carrascosa, V. Julian, Tool and integrated application development environment. Document identifier: AT/2011/D6.2.3/v1.0 Project: CSD2007-0022, INGENIO 2010 (2011), http://www.agreement-technologies.org/
99. A. Rotolo, L. van der Torre, Rules, agents and norms: guidelines for rule-based normative multi-agent systems. RuleML Eur. **6826**, 52–66 (2011)

100. S. Rougemaille, F. Migeon, C. Maurel, M.-P. Gleizes, Model driven engineering for designing adaptive multi-agents systems. in *Engineering Societies in the Agents World VIII: 8th International Workshop, ESAW 2007, Revised Selected Papers* (Springer-Verlag, Berlin, 2008), pp. 318–332

101. M. Saeki, H. Kaiya, Supporting the elicitation of requirements compliant with regulations. in *CAiSE '08* (2008), pp. 228–242

102. J.W. Schemm, C. Legner, H. Sterle, E-contracting: Towards Electronic Collaboration Processes in Contract Management (2006), pp. 255–272

103. V. Seidita, M. Cossentino, S. Gaglio, Using and extending the spem specifications to represent agent oriented methodologies. in *Agent Oriented Software Engineering (AOSE)* (2008)

104. A. Siena, J. Mylopoulos, A. Perini, A. Susi, Designing law-compliant software requirements. in *International Conference on Conceptual Modeling, ER '09* (2009), pp. 472–486

105. E. Solaiman, C. Molina-Jimenez, S. Shrivastav, Model checking correctness properties of electronic contracts, *Service-Oriented Computing—ICSOC*. LNCS, vol 2910 (Springer, Berlin, 2003), pp. 303–318

106. R. Soley, OMG Staff Strategy Group. Model driven architecture (2000), ftp://ftp.omg.org/pub/docs/omg/00-11-05.pdf.

107. A. Sturm, O. Shehory, A framework for evaluating agent-oriented methodologies, in *Agent-Oriented Information Systems*. LNCS, vol. 3030, ed. by P. Giorgini, B. Henderson-Sellers, M. Winikoff (Springer, Berlin, 2004), pp. 94–109

108. P. Telang, M. Singh, Specifying and verifying cross-organizational business models: An agent-oriented approach. IEEE Trans. Serv. Comput. **5**(3), 305–318 (2012)

109. P.R. Telang, M.P. Singh, Enhancing Tropos with Commitments, in *Conceptual Modeling: Foundations and Applications* (Springer, Berlin, 2009), pp. 417–435

110. P.R. Telang, M.P. Singh, Comma: A commitment-based business modeling methodology and its empirical evaluation. in International Conference on Autonomous Agents and MultiAgent Systems (IFAAMAS, 2012), pp. 1073–1080

111. P.R. Telang, M.P. Singh, Comma: A commitment-based business modeling methodology and its empirical evaluation. in *Proceedings 11th International Conference on Autonomous Agents and MultiAgent Systems* (2012)

112. Q.-N. Tran, G. Low, Comparison of ten agent-oriented methodologies. in *Agent-Oriented Methodologies*, ed. by B. Henderson-Sellers, P. Giorgini (Idea Group Publishing, London, 2005), pp. 341–367

113. I. Trencansky, R. Cervenka, Agent modelling language (AML): a comprehensive approach to modelling mas. Informatica **29**(4), 391–400 (2005)

114. A. Van Dijk, Contracting workflows and protocol patterns. in *International Conference on Business Process Management* (2003), pp. 152–167

115. J. Vázquez-Salceda, R. Confalonieri, I. Gomez, P. Storms, S.P. Nick Kuijpers, S. Alvarez, Modelling contractually-bounded interactions in the car insurance domain. *DIGIBIZ* (2009)

116. F. Vigano, M. Colombetti, Specification and verification of institutions through status functions. in *Coordination, Organizations, Institutions, and Norms in Agent Systems II*, ed. by P. Noriega, J. Vzquez-Salceda, G. Boella, O. Boissier, V. Dignum, N. Fornara, E. Matson, LNCS, vol 4386 (2007), pp. 115–129

117. F. Viganò, M. Colombetti, Symbolic model checking of institutions. in *International Conference on Electronic Commerce* (2007), pp. 35–44

118. C.D. Walton, Verifiable agent dialogues. J. Appl. Logic **5**(2), 197–213 (2007) (Logic-Based Agent Verification)

119. M. Wooldridge, P. Ciancarini, Agent-Oriented Software Engineering: The State of the Art. in Agent-Oriented Software Engineering: First International Workshop. LNCS, vol 1957 (Springer, Berlin, 2001), pp. 55–82

120. M. Wooldridge, M. Fisher, M. Huget, S. Parsons, Model checking multi-agent systems with mable. in International Conference on Autonomous Agents and MultiAgent Systems (ACM, New York, 2002), pp. 952–959

121. M. Wooldridge, M. Fisher, M.-P. Huget, S. Parsons, Model checking multi-agent systems with mable. in *International Joint Conference on Autonomous Agents and Multiagent Systems* (ACM, New York, NY, USA, 2002) pp. 952–959

122. F. Zambonelli, N.R. Jennings, M. Wooldridge, Organisational rules as an abstraction for the analysis and design of multi-agent systems. Int. J. Softw. Eng. Knowl. Eng. **11**(3), 303–328 (2001)

123. F. Zambonelli, N.R. Jennings, M. Wooldridge, Developing multiagent systems: the gaia methodology. in ACM Transactions on Software Engineering Methodology, vol 12 (Springer-Verlag, New York, 2003), pp. 317–370

124. I. Zinnikus, C. Hahn, K. Fischer, A model-driven, agent-based approach for the integration of services into a collaborative business process. in International Conference on Autonomous Agents and MultiAgent Systems (2008)

Printed in the United States
By Bookmasters